The School Food Revolution

The School Food Revolution

Public Food and the Challenge of Sustainable Development

Kevin Morgan and Roberta Sonnino

London • Washington, DC

First published in hardback by Earthscan in the UK and USA in 2008
Published in paperback in 2010

Earthscan Ltd, Dunstan House, 14a St Cross Street, London EC1N 8XA, UK
Earthscan LLC,1616 P Street, NW, Washington, DC 20036, USA
Earthscan publishes in association with the International Institute for Environment and Development

For more information on Earthscan publications, see www.earthscan.co.uk or write to earthinfo@earthscan.co.uk

ISBN: 978-1-84407-482-2 hardback
ISBN: 978-1-84971-086-2 paperback

Typeset by FiSH Books, Enfield
Cover design by Susanne Harris

A catalogue record for this book is available from the British Library

Library of Congress Cataloging-in-Publication Data

Morgan, Kevin.
The school food revolution : public food and the challenge of sustainable development / by Kevin Morgan and Roberta Sonnino.
p. cm.
ISBN 978-1-84407-482-2 (hardback)
1. School children—Food. 2. School children—Nutrition. I. Sonnino, Roberta.
II. Title.
LB3475.M67 2008
371.7'16—dc22

2008019749

At Earthscan we strive to minimize our environmental impacts and carbon footprint through reducing waste, recycling and offsetting our CO_2 emissions, including those created through publication of this book. For more details of our environmental policy, see www.earthscan.co.uk.

Mixed Sources
Product group from well-managed forests and other controlled sources
www.fsc.org Cert no. SGS-COC-2482
© 1996 Forest Stewardship Council

Printed and bound in the UK by TJ International, Padstow. The paper used is FSC certified.

For school food reformers –
cooks and caterers, governors and teachers,
buyers and suppliers, parents and politicians –
who are striving to change the way our children eat

Contents

List of Boxes, Figures and Tables

Boxes

Figures

Tables

Acknowledgements

Research for this book has been sponsored by the UK Economic and Social Research Council (RCTO373). We are very grateful for all the support that ESRC has given us to conduct research in Europe and the US. We would also like to thank the British Academy, the World Food Programme and the Bill and Melinda Gates Foundation for sponsoring our research on school feeding in developing countries.

A comparative study generates many debts, and this book is no exception. We have benefited enormously from the help and support of many people who have taken time out of busy schedules to meet with us, answer our questions and provide critical feedback as we sought to understand the harsh realities of the school food service in different countries. In the academic world, we would like to especially thank our colleagues Terry Marsden, Mara Miele, Yoko Kanemasu and Adrian Morley (Cardiff University) for their contributions to our research on school meals over the years. Tanja Bastia and Tricia Sexton were also part of the research team before they left for pastures new. Richard Cowell (Cardiff University), Gianluca Brunori (University of Pisa), Tim Lang (City University, London), Andrew Sayer (Lancaster University) and Sergio Schneider (University of Rio Grande, Brazil) have all provided very valuable comments on early drafts of various chapters (though they bear no responsibility for the content of the book, of course). Janice Edwards, Andrew Edwards and Matthew Leismeier (Cardiff University, School of City and Regional Planning) have done an outstanding job on preparing many of the figures and tables in the book. Special thanks also to Denise Phillips for her invaluable help at the closing stage of the writing. Janet Poppendieck (Hunter College) deserves a special mention for that three-hour phone conversation that helped us to understand the historical and social context of school lunches in the US. Many thanks also to Toni Liquori (Columbia University) for inviting us to talk about our research in the US and for introducing us to some of the key players in the New York City school meals system. We are also extremely grateful for the patience and local knowledge of Marion Kalb (Farm-to-School Program).

Without the expert help of the professionals who manage the school meals systems we studied, this book would not have been possible. In New York, we would like to thank David Berkowitz, the Executive Director of School Food, and Jorge Collazo (Head Chef), for their time and their expertise. Many thanks also to Angie Dykshorn, whose research support has been first class, and to Latoya Spruill for her fantastic hospitality. In Rome, we owe a great debt of gratitude to Maria Coscia, the former Councillor for Education, and to Silvana Sari, the Director of the Education Department, a real leader in the school food revolution. Paolo Agostini, Patrizia Rani and Luisa Massimiani have also been extremely helpful and cooperative during our research. Finally, thanks to Paola Trionfi (Associazione Italiana per l'Agricoltura Biologica) for arranging our interviews at the national level in Italy and to Sara Cusatelli Lener for organizing the press conference with former Mayor Walter Veltroni and for helping us to include the drawing made by a Roman child on the cover of the book. In the UK, we thank Teresa Filipponi (Welsh Local Government Association) for her enthusiastic help. Robin Gourlay (East Ayrshire Council) has welcomed us with unforgettable warmth and has remained a crucial point of reference for us throughout our research. Many thanks also to Kay Knight and Peter Cook (South Gloucestershire County Council), to Elin Cullen (Carmarthenshire County Council), to Bobbie Bremerkamp (London Borough of Greenwich) and to Claire Pritchard (Greenwich Cooperative Development Agency), all of whom gave a generous amount of time to our research.

We also appreciate the interest that the media have demonstrated in our research. Roberta Grasselli and her husband Davide took the trouble to come all the way to Cardiff to film an interview with us for their documentary on school lunches. Thank you also to the BBC, which has broadcast our research on school food at various stages with great professionalism.

Earthscan has provided invaluable technical support. We would like to thank in particular Alison Kuznets, Olivia Woodward, Hamish Ironside and Tim Hardwick for their commitment and their availability.

Finally, we are also indebted to family and friends for putting up with us at the most critical stages of the research and writing. Roberta would like to especially thank her parents, Lori and Sergio, and her sister Annalisa, all of whom have provided very special help and support during the good as well as the difficult times that a long research process inevitably entails. Kevin would like to thank his family, Sue, Louis and Robin, for their emotional support and their political commitment to school food reform, and Doreen Taylor, his mother-in-law, whose paper clipping service made her a de facto member of the research team.

Introduction

At first sight, the idea of serving fresh, locally produced food in schools looks very simple. But nothing could be further from the truth. One of the chief aims of this book is to explain why the locally sourced school meal, such a simple confection in theory, turns out to be surprisingly complex in practice. So complex, in fact, that it remains a daunting challenge for everyone involved in the school food chain, especially dinner ladies, caterers, procurement managers, suppliers, regulators and parents. Part of the explanation, we believe, has to do with the fact that in many countries, particularly the UK and the US, the idea runs against the grain of some very powerful cultural conventions – like the notion that there is nothing special about food; that food is just one industry among others; that cost takes precedence over quality in public sector catering; and that the provenance of food is a matter for the exclusive restaurant, rather than the school canteen. As we will see, the school food revolution challenges each of these conventions.

Before we preview the arguments of the book, we should perhaps explain how and why we got involved in school food provisioning. The research grew out of a local food project in Powys, a rural county in Wales, where a group of people had become so concerned about the modern food system – and its adverse effects on health, local economies and the global environment – that they decided to try to change it. The Powys Food Links project came into being in 2000 with what seemed to be a simple and unpretentious aim: to get locally produced food into a local hospital. The people involved hoped and believed that their project would benefit patients and producers alike: the former from eating fresh and nutritious food, the latter from finding a new market.

However, the seemingly simple aim of the Powys Food Links project was frustrated at every turn. Indeed, everything seemed designed to prevent locally produced food from being consumed locally. Time and again, the same barriers kept cropping up: European Union procurement rules that appeared to be stacked against local food transactions; UK local government regulations that construed 'value for money' in very narrow terms; complex tendering procedures that seemed designed to exclude small producers; catering customs that favoured big national suppliers; and the audit culture in the National Health Service, which found it virtually impossible to account for the health gains of nutritious food.

Any one of these barriers might have been enough to prevent a local food chain taking root, but together they totally overwhelmed the local experiment. Having followed this sad catalogue of events from afar, we approached Powys Food Links to discuss the lessons of their failure, not least because we believed that this local project had a global resonance.

Along with the Soil Association, the premier certification body for organic produce in the UK, the School of City and Regional Planning at Cardiff University joined the local campaigners to launch a new venture – the Powys Public Procurement Partnership (4P) project – to examine the barriers to local food procurement. One of the components of the project was a study of public procurement regulations in the UK food chain, with a special focus on schools and hospitals, which have the greatest need of nutritious food. Although this was a very modest study – lack of funding meant that it had to be undertaken as a labour of love – the report attracted more attention than we ever imagined, especially from the worlds of policy and practice (Morgan and Morley, 2002). The main reason why the report resonated was a powerful combination of time and place: the topic was simply part of the zeitgeist sweeping through the UK's food and farming sector, where politicians were beginning to show a new interest in the purchase of healthier food in schools as a result of the moral panic about childhood obesity.

Although the report was largely well received, it had its critics. Members of the Local Authority Caterers Association (LACA), for example, took against it in a big way, accusing us of undermining the school food system by criticizing the quality of the food. By far the most difficult moment came when one school caterer, a woman who had been incredibly forthcoming in interview, asked us to remove the financial data she had given us on what local authorities were actually spending on food ingredients. Having been sent a draft version of the report, she knew there was no mention of her or her local authority, so there was no ethical obligation on our part to censor the data. To our knowledge, the data had never been released into the public realm before, and we felt that parents and politicians had to be made aware of the reality of the cheap food culture in schools. The controversial section of the report said:

> *Catering staff are being asked to perform minor miracles daily when they try to deliver quality meals at a price which most people would consider impossibly low – between 32 and 38 pence for a primary school meal. One local authority catering manager said that the cost pressures spawned by the compulsory competitive tendering revolution had 'driven us down to the bone on cost', so much so that she could not afford to purchase more local food if it cost more.* (Morgan and Morley, 2002, p50).

We decided not to censor the above section because we thought it would help to raise political awareness about the woefully inadequate level of funding for school food – along with the derisory terms and conditions of the cooks and caterers. The LACA eventually accepted that, like them, we had the best interests of the school food service at heart and subsequently invited us to address their annual conference on three consecutive occasions.

Working on the 4P project convinced us that, as well as being an intrinsically significant research topic in its own right, the school food service was also a prism through which we could explore some of the most compelling questions that face us today. How can the public realm be encouraged to set more demanding standards for health and wellbeing, particularly for the poorest in society? How can public procurement become a creative force for sustainable development rather than being stymied by (real and imagined) regulations? And how can the school food service play its part in combating the obesogenic environment in which children are reared?

Raising serious questions about the school food service induced a curious response from fellow British academics: an amalgam of mirth and bewilderment that we have not encountered in other countries. This curious response may have something to do with a cultural stereotype that has always sought to present the British school meal in a comical light. For generations, school meals in Britain were portrayed in film and literature as something to be endured rather than enjoyed, a character-forming rite of passage and, therefore, not something to be taken too seriously. Fortunately, this cultural stereotype has recently been banished in the UK, where the humble school meal is now perceived to be a litmus test of the government's political commitment to sustainable development (Morgan, 2004a).

Research funding also helped to establish the school food service as a legitimate object of study, and we are truly grateful to the Economic and Social Research Council for providing the means for us to study school food reform in comparative perspective – specifically in the UK, Italy and the US. We researched school food reform both in large cities – Rome, London and New York – and in smaller rural settings to understand the effects of cultural diversity and spatial scale on the design and delivery of the service. London and New York were chosen because they are the premier cities in their respective countries – world cities par excellence – and were both trying to reform their school food services in a highly diverse cultural context and on a very large spatial scale. Rome was chosen because it was reputed to have one of the best school food services in Europe, a reputation that stood the test of taste. Following an invitation from the Mayor of Rome, who had arranged a joint press conference to discuss the Roman model, we sampled a typical school lunch in the Rio de Janeiro Primary School. The experience confirmed the evidence we were getting from other, more formal sources that the City of Rome takes its school food service very seriously.

Additional research funding from two other sources – the British Academy and the Bill and Melinda Gates Foundation – enabled us to extend our work into the developing country context. The Gates Foundation had decided to fund a new World Food Programme initiative called *Home-Grown School Feeding*, which aimed to use school feeding programmes to purchase locally produced food rather than imported food, creating new markets for small producers in the process. The WFP commissioned us to review the scope for home-grown school feeding in selected developing countries, particularly with reference to public procurement.

Having explained the origins of the book, let us briefly introduce the main lines of argument in the chapters that follow. Chapter 1 situates school food reform in the theoretical debate about sustainable development, the appropriate context in which to understand the larger issues at stake. We argue that the ambiguity surrounding sustainable development is inherent in the concept itself because, as a normative standard like democracy and justice, it acquires meaning and substance only in concrete socio-cultural contexts. Generally speaking, sustainable development is informed by three fundamental principles: more equitable forms of economic development across space and time, more participatory democratic structures and greater environmental integrity.

Drawing on these principles, we try to operationalize the concept of sustainable development by examining one of the most contentious issues in agri-food studies – the *re-localization* of the food chain. Also in Chapter 1, we examine the claim that a more localized food system is necessarily a more sustainable food system, an idea that is contested by the emerging literature on the 'local trap', which suggests that it is wrong to ascribe social values to spatial scales, since the latter have no inherent qualities.

Far from being a purely theoretical debate, the question of what spatial scales are most conducive to sustainable development is also an intensely political one, since the state is now beginning to use its power of purchase to promote more sustainable food systems. This is especially the case where the state takes its commitments to sustainable development seriously – where, in other words, there is a *Green State*. The Green State is, in our view, more an ideal type than an actually existing state; a vision of what might be, rather than what is; a political process of becoming, rather than arriving. Chapter 1 concludes by examining how a Green State could expedite school food reform.

In Chapter 2 we explore the Byzantine world of public procurement. Along with the powers to tax and regulate, the *power of purchase* is one of the most influential means through which the state can effect behavioural change in economy and society. But the world of public procurement presents us with a curious paradox: in most countries, its economic significance is strangely out of sync with its political status. Despite its potential power, in other words, the story of public procurement is largely a tale of untapped potential. To try to explain this paradox, we explore the trials and tribulations of the UK, which is aspiring to be a world leader in procurement – having been something of a laggard in this regard for quite a long time.

The barriers to creative or sustainable procurement in the UK are largely the same as they are elsewhere, and they include such things as cost perceptions, poor knowledge, lack of leadership, organizational inertia and legal uncertainty as to what can be done under the existing regulations. Public procurement managers are not free agents when it comes to purchasing goods and services, particularly in the highly regulated world of food. These regulations embrace the international rules of the World Trade Organization (WTO), where the overriding aims are to deregulate trade and to ensure that public contracts are subject to open competition. These aims are reinforced by EU and US regulations, wherein public bodies are also required to secure 'best value', a concept that can be interpreted narrowly or broadly, depending on the national interplay of culture and politics.

One of the most controversial issues in the EU and the US is whether public bodies, particularly sub-national governments and school districts, have the legal right to specify local food in their contracts. Here we try to explore the myths that surround the issue in Europe, where, for example, UK procurement managers have convinced themselves that EU regulations prohibit specifying the use of local food in public contracts. While it is indeed illegal to specify local products that can only be supplied by local producers (a stance that contravenes the EU principle of non-discrimination), it is possible to specify for produce such qualities as fresh, seasonal, organic and certified – qualities that are used to secure local produce in all but name in countries like Italy, for example.

The US is also struggling with the legalities of local food procurement. While the US Department of Agriculture (USDA) maintains that federal procurement regulations prohibit specifying state or local geographic preferences, other legal experts claim that local food procurement is legal because it is 'not disallowed'. Regulatory confusion is now the biggest single barrier to the use of locally produced food in American public schools.

We conclude Chapter 2 by arguing that public procurement could help to bring the Green State a little closer – if its potential was more clearly recognized, if public sector managers had the competence to design more creative tenders, and if the rules and regulations were to acknowledge that fresh food, being vital to human health and wellbeing, need not be just another product under the WTO regime.

With the first of three urban case studies, Chapter 3 explores the world of school food reform in New York, the most populated, ethnically mixed and socially diverse city in the US. No country in the world has allowed the fast food industry to colonize the school environment to a greater extent than the US, with the result that the American system has regressed since the National

School Lunch Act (NSLA) was approved in 1946. One of the original aims of the NSLA was 'to safeguard the health and well being of the Nation's children', a public ethic of care that atrophied in the final quarter of the 20th century. If the original reason for the NSLA was that children were not getting enough food, the concern today is that they are getting too much of the wrong food, with nearly a quarter of New York's children aged 6–11 classified as obese, against a national average of 15 per cent. In this challenging context, New York City's school food reform is trying to focus on one overriding issue: better nutrition.

One of the most controversial aspects of New York's reform strategy is the fact that healthy school meals are actually disguised as fast food to make them more attractive to students. Some food radicals are highly critical of this strategy, claiming that it imports the neo-liberal values of the market into the public school system. We adopt a more pragmatic approach because we believe that the city's school food reform is beginning to show some promising results, such as boosting the nutritional value of school food, enhancing student participation in the school lunch programme and creating new opportunities for local food procurement. These are 'little victories' to be sure, and they will need more federal support to be sustained, but they are not insignificant in the world's premier fast food nation.

Chapter 4 examines the most successful example of school food reform in the book, the City of Rome's 'quality revolution' that began in 2001. Rome can draw on a rich national food culture, but this does not explain why it has been one of the most innovative cities in Italy on the school food front, where it has banned GM foods and most frozen vegetables and replaced them with new, nutritionally balanced meals using fresh, seasonal and organic fruit and vegetables. At the same time, Rome has also upgraded the environmental quality of its school food service. In the bidding process, catering companies have been rewarded for their environmental credentials, including the provision of organic food, and, more recently, they have even been encouraged to source products from 'bio-dedicated' food chains, which means foods that have been produced, processed, packaged and distributed by companies that operate exclusively in the organic sector.

With strong political backing from the Mayor, the 'quality revolution' has been spearheaded by a highly committed team of public-spirited officials in the city's education department. One of the award criteria they have designed is called 'guaranteed freshness' and promotes food products that were harvested no more than three days before being served in the schools. To assess the freshness of the foods offered, and in an effort to reduce the pollution and transport costs of food, Rome is the first city in the world to explicitly look at food miles – specifically, the number of miles and hours that separate harvesting and consumption. These reforms have been underwritten by a significant investment by the city, which is perhaps the real index of Rome's political commitment to healthy eating and social justice.

The local significance of a strong mayor is also apparent in Chapter 5, which explores London's plans to become a sustainable world city. One of these plans, a healthy and sustainable food strategy for the capital, is predicated on a very simple observation: that the food system, as it currently exists, threatens to undermine London's ambition to become 'a world-class, sustainable city'. Although London was recently crowned the 'gastronomic capital of the world' on account of its rich restaurant culture, other aspects of its food system leave much to be desired, with childhood obesity more prevalent in London than elsewhere, many people struggling to access affordable, nutritious food and school food reform very uneven across the capital's 33 boroughs.

London may pride itself on being a world city, but it remains part of the British school food system, which was allowed to atrophy in the same way as the US system, though not to the same extent. The London Mayor has very little power to counteract this trend, because with school food it is the boroughs that control the service, not the Mayor's Office. Since the borough is the key unit of analysis, we include a case study of Greenwich to illustrate the scope and limits of school food reform in London.

Chapter 6 shifts the focus from large urban centres to small rural areas. We selected three local authority areas – Carmarthenshire in Wales, East Ayrshire in Scotland and South Gloucestershire in England – because they were acknowledged to be the pioneers of school food reform in their respective parts of the UK. Each of them can be used to highlight an important aspect of school food reform. What we see in the case of Carmarthenshire, for example, is a small rural county that was ahead of its time. It was criticized by regulators for having a 'high-cost' school food service before healthy eating became fashionable, a criticism that used the 'cheap food' catering culture of other local authorities as the benchmark. The fact that the county disputed this assessment, drawing on community health arguments to do so, is a tribute to its local leadership.

East Ayrshire has been highly innovative too, especially with respect to its procurement strategy. The Scottish council redesigned its bidding process to enable local and organic suppliers to enter the market. The contract was divided into as many as nine lots to help small suppliers to compete with large national suppliers. To improve the ingredients, award criteria were equally based on price and quality, an approach that rewarded suppliers' proposed timescale from harvest to delivery, their ability to supply Fair Trade, seasonal and ethnic foods, their contribution to biodiversity and their compliance with animal welfare standards. In 2006, 50 per cent of the ingredients in the primary school meals were organic, about 70 per cent were locally sourced and more than 90 per cent of all food on the school menu was unprocessed.

South Gloucestershire adds another dimension to the story of school food reform, highlighting the role of labour in the delivery of the service. The head of the service is strongly committed to the view that economic and environmental sustainability cannot be achieved without a well-trained and highly motivated catering staff. For this reason, kitchen staff in this rural county underwent an intensive training programme on nutrition and customer care to enable them to deliver a more healthful service – a move that was also meant to highlight the dignity of labour.

In Chapter 7 we enter the profoundly important, but little known, world of school feeding in developing countries. Although a single chapter cannot hope to do justice to the noxious cocktail of hunger, disease and underdevelopment in the poorest countries of the world, we try to analyse the problem from a different angle by focusing the discussion on the challenge of *home-grown* school feeding. Conventional school feeding is largely based on food imports from abroad, a mechanism that can damage the indigenous agricultural sector in developing countries, even though the motive may be a humanitarian one. As noted earlier, the chief aim of the home-grown school feeding programme is to create markets for local producers in the process of promoting the health and education of the children involved.

Our key argument, however, is that home-grown school feeding is about so much more than school food. It is about fashioning a robust and transparent framework for collective action. It is about creating, and sustaining, a dedicated budget to enable the system to survive the vagaries of the electoral cycle. It is about learning to deploy the power of purchase in a way that nurtures small-scale farmers, helping them to make the transition from subsistence farming to commercial agriculture. And it is also about keeping corruption at bay within the state and enlisting the active support of civil society and the business and donor communities. In short, the home-grown model is the entire drama of development in microcosm. Consequently, it needs to be understood as a learning-by-doing exercise in which the end product, the provision of nutritious food, is just one part of a much larger process.

In Chapter 8 we try to relate school food reform to some of the largest questions that societies can ask themselves in the 21st century. We argue that school food reform needs to be protected and promoted by a new *public* ethic of care, because caring, as a moral principle and as a social activity, has been largely confined to the private realm, where the burden mainly falls on women. We explore the prospects for a new public ethic of care in two arenas. First, in the sectoral context of school food reform, where we outline what a sustainable system might look like. And, second, in the societal context of obesity and malnutrition, where we argue that these twin crises cannot be solved without more concerted collective action in the public realm.

The book concludes on a speculative note by exploring the prospects of a transition from school food provisioning to community food planning. Community food planning could help to extend the public plate to new social and spatial scales, enabling it to serve adults as well as children and helping the state to honour the most basic human right of all – the right to food. Indeed, the fact that chronic hunger affects so many people in the 21st century makes a mockery of a 'right' that was first proclaimed 60 years ago, when in 1948 the UN first recognized the right to food in the Universal Declaration of Human Rights. Of all the ethical obligations that societies accept when they commit themselves to sustainable development, none is surely more important than eradicating chronic hunger. Meeting this challenge is about political will, not economic resources. The World Food Programme has estimated that it will take US$1.2 billion a year to ensure that no child goes to school hungry in Africa. As we argue in Chapter 8, this amounts to roughly one per cent of what the US government was annually spending on the Iraq war up to 2007. Although hope springs eternal, as the saying goes, it is difficult to be hopeful that the right to food will ever be honoured unless we shift our priorities from warfare to welfare.

A new set of priorities will also be needed to realize the potential of what we call the school food revolution. For the most part, this is a *conceptual* rather than an actual revolution, more aspiration than reality. Governments around the world now accept that an investment in school food today is an investment in the health and welfare of their citizens tomorrow; with notable exceptions like Finland and Sweden, however, they continue to treat it as a commercial service – a case of willing the ends but not the means. The school food revolution will have realized its potential when the real nature of the service is fully understood and when good food on a child's plate is the norm, not the exception.

1

Public Food and Sustainable Development: Barriers and Opportunities

Sustainable Development: From Theory to Practice

At the dawn of the 21st century, the world is confronting a global environmental crisis of unprecedented magnitude and reach. The crisis, in itself, is not new: human interaction with the natural world has long damaged our environment. What is new is the speed at which pollution levels increase, the Earth's temperature rises and animal and plant species disappear for ever.

Also relatively new is the widespread awareness that these global environmental problems are endangering not just our ecosystems; they are also threatening the future of our economies. At the end of 2006, Nicholas Stern, a former chief economist at the World Bank, published a seminal report arguing that global warming could deliver an economic blow of 5–20 per cent of GDP to world economies because of natural disasters and the creation of hundreds of millions of refugees displaced by droughts or rising sea levels. Dealing with climate change now, the Stern Report contends, would cost just one per cent of the world's GDP; but if the problem is not tackled within a decade, we will be forced to invest almost US$1000 for every person on the planet – a figure that could push the global economy into its worst recession in recent history (Stern, 2006). Although the data provided by the Stern Report are still debated and discussed, the impact of this document in the media all over the world shows that global environmental problems have now entered policy debates in all fields and at all levels.

For many experts and policymakers, this global and multifaceted environmental crisis is raising the need to rethink the concept of development, shifting its fundamental goal from the basic idea of quantitative growth to the more encompassing notion of qualitative improvement in people's lives (Daly, 1996). Practically, this means devising development strategies that move beyond the old modernization paradigm, with its narrow focus on economic growth, to embrace also the environmental and social dimensions of our lives. In simple terms, it means promoting a development model that emphasizes, rather than undermines, the interdependence of economy, society and nature. In this context, the concept of *sustainable development* has become the most powerful ideological tool to catalyse attention on the social and ecological conditions necessary to support human life at a certain level of wellbeing through future generations (Earth Council, 1994).

It has been 20 years since the Brundtland Report, which provided the first celebrated definition of sustainable development as 'development that meets the needs of the present without compromising the ability of future generations to meet their own needs' (WCED, 1987). Despite the popularity of the concept in both academic and political discourse, so far progress towards sustainable development has overall been 'slow, piecemeal and insubstantial' (Carter, 2007, p356). With the exception of local debates around Local Agenda 21,[1] the discussion around sustainability has taken place primarily at a global and theoretical level, generating endless and mostly abstract speculations over the exact meaning of sustainable development.

Some scholars have emphasized a specific dimension of sustainability. For example, from a strictly economic perspective the goal of sustainable development is to ensure that the per capita income of future generations will be no less than that of current generations (Tisdell, 1999a, p24). This 'weak' version of sustainable development accepts a commitment, where possible, to protect natural resources, but it rejects the idea that economic activity should be confined within predetermined environmental limits (Jacobs, 1999, p31). Environmentalists have opposed to this anthropocentric view of sustainability a more ecocentric perspective, which concentrates on biodiversity and the protection of natural resources (Gibbon and Jakobbson, 1999, p106). This 'strong' version of sustainable development emphasizes the notion of 'environmental limits' – that is, it is based on a commitment to living within the limits created by the 'carrying capacities' of the environment – and situates humankind *in* nature and not *above* it (Jacobs, 1999, p31). Still other scholars have focused on social sustainability, which implies generating enough wealth for a society to reproduce itself, to maintain its institutions and to provide a sense of cohesion and community for its members (Gibbon and Jakobsson, 1999, p107).

On other occasions, the focus of the debate on sustainable development has been multidimensional. In an effort to promote a more integrated development approach, some scholars equally emphasize economic, social and environmental goals (Tisdell, 1999b). Pretty (1999), for instance, argues that sustainable systems must accumulate stocks of five different types of capital: natural capital (nature's goods and services), social capital (the cohesiveness of people in their societies), human capital (the status of individuals), physical capital (local infrastructure) and financial capital (stocks of money).

In the light of these fundamental differences in how the goals of sustainability have so far been interpreted, some have argued that sustainable development is still a vague, if not ambiguous, concept, perhaps 'too unrealistic, biased and naïve' (Gibbon and Jakobsson, 1999, p104). For some scholars, sustainable development holds the potential to fit in with a wide variety of different political and economic agendas; hence, it can become 'an insubstantial and clichéd platitude unworthy of further interest or research' (Drummond and Marsden, 1999, p1). Richardson (1997, p43) captures this view by arguing that sustainable development is:

> *a political fudge: a convenient form of words [...] which is sufficiently vague to allow conflicting parties, factions and interests to adhere to it without losing credibility. It is an expression of political correctness which seeks to bridge the unbridgeable divide between the anthropocentric and biocentric approaches to politics.*

In our view, this is an essentially flawed debate, which fails to recognize that sustainable development has a relative and not an absolute meaning. Like many other important concepts, such as democracy and justice, it is fundamentally a 'normative standard that serves as a meta-objective for policy' (Meadowcroft, 2007, p307). At the theoretical level, sustainable development can contribute to shaping and focusing political and scientific debates; in this sense, its importance lies in what it symbolizes (Richardson 1997, p53). However, in more practical terms, it is crucial to remember that different societies differently construct and value their environments (Redclift, 1997). For this simple reason, sustainable development acquires concrete meanings and substance only in each specific environmental and socio-cultural context. Hence, the implementation of sustainable development requires going beyond the search for a universal formula to 'embrace a plurality of approaches [...] and perspectives [...], accept multiple interpretations and practices associated with an evolving concept of "development", and support a further opening up of local-to-global public spaces' (Sneddon et al, 2006, p254). In short, implementing sustainable development is fundamentally a matter of negotiating its normative principles and adjusting them to contextually dependent priorities and needs.

Broadly speaking, there are three fundamental principles that inform sustainable development as a normative standard. First, sustainable development is about promoting more equitable forms of *economic development* across space and time. In other words, it is about attempting to meet the basic needs of *all* human beings while also recognizing the potential for imposing risks or costs on future generations.

The concept of *sustainable consumption* has emerged as a key tool to achieve this goal. Today it is widely recognized that there are growing disparities between mass consumption patterns in the world's North and the inability to meet basic consumption needs in the South. To give just one example, Chasek et al note (2006, p3) that we would need about US$13 billion a year to provide basic healthcare and nutrition for the world's poor; in Europe and the US, people spend about US$17 billion a year on pet food. By redefining notions of 'wealth', 'prosperity' and 'progress', sustainable consumption initiatives attempt to construct new social and economic institutions for governance that value the socio-environmental aspects of wellbeing alongside the economic dimension (Seyfang, 2006, p385). Fair Trade labels provide a good example of this kind of initiative. In fact, their aim is to reduce the direct impact of Northern consumption on scarce resources, while at the same time improving the social and economic condition of Southern communities that supply these resources (Carter, 2007, p219). In the context of current efforts to implement more equitable socioeconomic systems, sustainable consumption has also become a goal within the wealthiest countries of the world, where the pressures of competitive spending are increasingly widening the gap between rich and poor, with major consequences in terms of social justice. For this reason, as Carter (2007, p220) explains, 'achieving sustainable consumption will [...] involve both an overall readjustment in the levels and patterns of consumption in rich countries and the provision of basic needs to the socially excluded poor'.

Second, sustainable development fosters *democracy* through a vision of interconnected and highly participatory communities. Stemming from a critique of liberal democracy and of its intrinsic hierarchies, bureaucracy, individualism and inequalities, this principle proposes a model of participatory democracy that produces more environmentally responsible governments and fosters greater individual autonomy and involvement (Carter, 2007, pp55–56; see also Eckersley, 2004). As we will discuss later in this chapter, the Green State has emerged as a powerful conceptual tool to describe the role and potential of environmentally engaged democratic institutions in promoting sustainable development.

Third, sustainable development is about *integrating environmental considerations* into our economic development strategies, under the assumption that effective environmental protection needs economic development and successful economic development depends on environmental protection. In this regard, it is important to point out that *environmental* considerations include more than just *ecological* issues. Sustainable development acknowledges that the environment also has a human dimension that encompasses the values, needs and aspirations of the people who inhabit it. Writing about the environmentally related nature of disease among poor people (especially children), von Schirnding (2002, p632) rightly points out that:

> *sustainable development cannot be achieved if there is a high prevalence of debilitating illness and poverty, and the health of populations cannot be maintained without healthy environments and intact life-support systems.*

The integration of environmental considerations into the political decision-making process then raises an urgent need to fully recognize the interdependence between the sustainability of the environment and the sustainability of the human species and to merge local strategies for reducing health inequalities with strategies for reducing environmental inequalities (Griffiths, 2006).

In this book, we will operationalize the concept of sustainable development by focusing on its three fundamental (and interrelated) principles of economic development, democracy and environmental integration. Many political and economic actors around the world have committed themselves to the ambitious agenda for change that these principles embody. In an effort to respond to the global environmental crisis, for example, many industrialized countries have produced national sustainable development strategies, often formally supported by the business world and civil society (Carter, 2007). However, much discussion is still taking place, in both scientific and policy circles, as to how to achieve in practice the radical shift in existing patterns of production and consumption that sustainable development requires.

Food has increasingly moved to the forefront of this debate. Compared to other sectors or industries, food has a unique status: we ingest its output (Morgan, 2007b). For this reason, it is a special prism through which to explore the interconnections amongst the economic, social and environmental dimensions of development. Indeed, food brings about a wide range of issues that lie at the heart of current sustainability debates: from public health to social inclusion; from sustainable consumption to the environmental implications of activities such as transport, processing and waste management. In many ways, food has then become a litmus test of our individual and collective commitment to sustainable development. In the remaining part of this chapter, we will explore the relationship between food and sustainable development. Our focus will not be restricted to the economic, ecological and social requirements necessary to sustain a food system over time. Rather, we will discuss the more general contributions that a carefully planned and managed food system can make to the implementation of the three broad principles of sustainable development.

Food and Sustainable Development: Re-localization as an Opportunity?

Social and natural scientists have shown over the years how food, in its most industrialized version, has widely contributed to the global environmental crisis – at all stages of the supply chain. At the *production* end, intensification of agriculture through an ever increasing use of pesticides and fertilizers has depleted the soil, polluted waters and killed wildlife. Agricultural specialization, in turn, has caused a dramatic decline in the number of crop varieties and a significant loss of biodiversity. As a result, landscapes, rural livelihoods and farming systems have all been progressively simplified (Pretty, 1999, p86). In the most industrialized countries, diverse and integrated farms employing local people have been replaced by specialized operations that rely on contract labour; processing facilities have become centralized and remote from many rural communities; and mechanization and the loss and consolidation of many farms (especially small farms) have reduced the need for human labour, with predictable negative effects on employment rates and rural service provision. In many areas of the developing world, these trends have irreversibly damaged not just local ecological systems – they have also negatively affected local sociocultural practices. For example, in countries like Mexico and India, intensified use of herbicides has killed non-crop plants that provided food, fodder and medicine to native peoples, thereby seriously threatening their cultural identity, traditions and survival strategies (McMichael, 2000, p27).

At the *manufacturing* stage of the food chain, the enormous amount of fossil fuel used to process and transport food has significantly contributed to global warming and pollution problems. In the US, research conducted at Cornell University showed that in the mid-1990s more than 100 billion gallons of oil were used every year to manufacture food. More recently, it has been calculated that the average food item in the US travels between 1500 and 2500 miles from farm to fork (Kaufman, 2005). The situation is even worse in the UK, where a recent estimate suggests that food items travel on average 5000 miles from field to table (Pretty et al, 2005). As the environmental base crumbles, and one of the main ingredients of industrial agriculture – cheap oil – disappears, the future of farming itself seems to be increasingly at risk (Sachs and Santarius, 2007, p10).

Turning to the *consumption* end of the food chain, it becomes evident how the global environmental problems associated with the industrialization of food are also raising challenges to food security and public health. As corporate farmers have expanded their operations across the globe to engage in commercial food production that supplies mostly unsustainable affluent diets, half a billion rural people have lost access to the land to grow their own food (McMichael, 2000, p27). In some areas of the world, the industrial food system has arguably attenuated income-related class differences in food consumption by democratizing access (Goodman, 2004, p13). In other areas, however, the displacement of rural economies has left behind a series of 'food deserts' where people – especially low-income people – have little or no access to fresh, nutritious and healthy food (Wrigley, 2002; Guy et al, 2004). At stake, then, is also our individual and social wellbeing. In the UK, in 2003 the costs of obesity plus overweight were estimated at £6.6–7.4 billion per year (House of Commons,

2004). In the US, the public health costs associated with seven diet-related health conditions have reached US$10 billion per year (Kaufman, 2005). Simultaneously, and almost ironically, the incidence of hunger and food insecurity is on the rise, not just in developing countries but also in the wealthiest nations of the world. For example, many American cities are increasingly unable to provide adequate quantities of food to those in need (Pothukuchi, 2004, p357).

For all these reasons, many counter-movements have looked at food as one of the best examples of the cultural reductionism and unsustainability of modernization and its corporate logic (McMichael, 2000, p27). However, as much as it has been accused in the past of contributing to the global environmental crisis, today food is increasingly seen as part of the solution. Indeed, scientists and policymakers alike are beginning to realize that food systems hold the potential to deliver the wider objectives of sustainable development – economic development, democracy and environmental integration. But heated debates are still taking place around the most basic question: *How* exactly can food contribute to the emergence of alternative socioeconomic systems that deliver the wider objectives of sustainable development?

For many scholars and activists, the contribution of food to sustainable development is inextricably linked to the implementation of re-localization strategies that increase local food production for local consumption. Re-localization, for example, lies at the core of the concept of *community food security*, which advocates food systems that strengthen localities and communities by creating spatially closer links among two or more food system activities (Pothukuchi, 2004; Feagan, 2007). The same idea is implicit in the notion of *food justice*, which Wekerle (2004, p385) relates to concepts such as 'resistance to globalization'. And, as we will see in Chapter 7, even traditional school feeding initiatives in developing countries are currently experiencing a *home-grown* revolution.

Scholars who have explicitly attempted to theorize the relationship between food and sustainable development have also centred their discussions around the idea of localization. For Kloppenburg et al (2000, p18), for instance, a sustainable food system involves production in a 'proximate system' that emphasizes 'locally grown food, regional trading associations, locally owned processing, local currency, and local control over politics and regulation'. Bellows and Hamm (2001, p281) identify two parameters to assess the impacts of local import substitution strategies for food production. These include local autonomy, or 'the political capacity of a diverse public to negotiate its food needs both locally and vis-à-vis non-local food system actors', and sustainable development, which they define as the ability of locally based food systems 'to contribute to the future integrity and health of human and non-human environments'.

In part, this celebration of 'the local' mirrors a broader tendency in social sciences to develop anti-globalization narratives that construct the global as 'hegemonic and oppressive' and the local as 'radical and subversive' (Born and Purcell, 2006, p200). Escobar (2001), for example, argues that the most powerful forces of globalization do not always have de-localizing effects. In some cases, they strengthen the boundaries of a place. For this reason:

> *it might be possible to approach the production of place and culture not only from the side of the global, but from that of the local; not from the perspective of its abandonment, but from that of its critical affirmation; not only according to the flight from places, whether voluntary or forced, but of the attachment to them.* (Escobar, 2001, pp147–148)

In this perspective, the local often emerges as the most appropriate scale to implement sustainable development. For Cavallaro and Dansero (1998, p37), for instance, 'the sustainability of development is an objective that should be territorialized in order to be pursued in practice, in that the carrying capacities and potential vary in each local context'. In a similar fashion, Curtis (2003, p83) argues that 'the road to environmental sustainability lies in the creation of eco-local economies, which are place-specific and bounded in space by limits of community, geography and the stewardship of nature'.

When it comes to food, however, there are also at least three more specific arguments used to support the view that local food systems hold the potential to deliver sustainable development (Born and Purcell, 2006). First, it is often assumed that local production is more ecologically benign and re-embeds farming into environmentally sustainable modes of production (Renting et al, 2003). There are two rationales behind this assumption. On the one hand, through the use of concepts such as 'foodshed' (Kloppenburg et al, 1996) and 'terroir' (Barham, 2003), it is emphasized that local 'natural' variables, such as micro-weather patterns and soil and water quality, have a positive influence on the quality of the food produced. On the other hand, re-localization is considered to be a very effective strategy to reduce 'food miles' and cut the energy and pollution costs associated with the transportation and distribution of food around the world.

Second, it is argued that, in contrast with the dis-embedding of economic transactions from socio-environmental contexts, which is inherent in the functioning of the globalized food system, local food chains promote 'socially embedded economies of place' (Seyfang, 2006, p386) that establish relationships of trust between producers and consumers, enhance social capital, create feedback mechanisms and strengthen local economic development. Feenstra clearly illustrates this type of argument when she writes (1997, p28) that 'the development of a local sustainable food system not only provides economic gains for a community but also fosters civic involvement, cooperation and healthy social relations'. O'Hara and Stagl (2001) also emphasize the virtues of social embeddedness and the opportunities offered by innovations like Community Supported Agriculture[2] to 'recover lost dimensions' of the market as a place of social interaction. In this context, especially in Europe, local food systems have been seen as contributing to the emergence of a new rural development paradigm, capable of resisting the conventional cost–price squeeze on agriculture through the development of new relationships and methods of adding value (Renting et al, 2003).

Third, local food is often assumed to be fresher, riper, more nutritious and thus healthier (Born and Purcell, 2006, p203; DeLind, 2006, p123). This view is especially popular amongst consumers from countries, like the UK, which have experienced a long series of food crises, often linked to the lack of traceability in the industrialized food system.

But is there a real connection between local food and sustainable development? Is re-localization the most effective strategy to design food systems that promote economic development, democracy and environmental integration? As we will see in the next section, another powerful discourse is emerging today. By questioning the assumption that food re-localization is key to sustainability, this discourse is raising new challenges for policymakers attempting to use food as a practical tool to close the gap between the rhetoric and reality of sustainable development.

Re-localizing the Food Chain: Sustainable Development or the Local Trap?

In the literature on food, the 'local', like 'sustainable development', is subject to many competing interpretations – so much so that most researchers now consider it as an extremely complex, if not problematic, concept, both theoretically and practically. In general, it has been demonstrated that alternative food practices (including local food practices) can easily be co-opted by the conventional sector (Sonnino and Marsden, 2006). This has happened, for example, in California, with the 'conventionalization' of the organic sector (Guthman, 2004), and also in Europe, where foods protected by labelling schemes (which were originally introduced to strengthen the relationship between a food product and its local context of production) often have to rely on international food supply chains to become and remain economically viable (Watts et al, 2005, p30).

While raising the theoretical need for new conceptualizations that account for the blurring of the boundaries between conventional and alternative food systems (Morgan et al, 2006; Sonnino and Marsden, 2006; Feagan, 2007), this research demonstrates that the actual meaning of food localization is highly contested and not always conducive to sustainability. In California, for example, 'local' food has many different, and at times conflicting, implications, which embrace protective and active forms of particularism as well as global ambitions (Allen et al, 2003). In Iowa, the local food movement is characterized, on the one hand, by a 'diversity-receptive localization' that embeds the local into a larger national or world community and recognizes that the content and interests of 'local' are 'relational and open to change'. On the other hand, however, there is also a 'defensive localization' that stresses the homogeneity of the 'local' and often defines itself in patriotic opposition to outside forces, thereby becoming 'elitist and reactionary' (Hinrichs, 2003, p37).

This literature warns of the 'perilous trap' of the local (Hinrichs, 2003). As Campbell (2004, p34) puts it, 'local social relationships, power relations and environmental management practices are not always positive, and communities can pursue elitist or narrow "defensive localization" strategies at the expenses of wider societal interests'. Protectionism, resistance to the 'other', the minimization of internal differences and separation (Feagan, 2007, p36) are all potential outcomes of these defensive forms of localization.

The 'local trap' idea has dismantled the assumption that local food systems unequivocally contribute to sustainable development – environmentally, socially, economically and nutritionally. With regard to environmental sustainability, Born and Purcell (2006) have explicitly warned against the danger of conflating scale with ecological goals by stressing, for instance, how consuming local corn in Iowa means consuming conventional capitalist agriculture. Moreover, they explain, in an arid state like Arizona, while the re-localization of the food system may well reduce the amount of fuel used for transport, at the same time it would require massive water inputs, thereby bringing ecological damage on other grounds. The example of the 'local' farmer selling 'local' milk produced with high inputs of nitrate fertilizer at an intensively managed farm in Devon is another case in point (Winter, 2003, p30). As Winter explains (2003, p31), this example reflects a general emphasis in the UK on a type of defensive localism, which emerged in the aftermath of the foot and mouth disease epidemic, whereby support for local farming has erased any concerns for environmental practices and agri-food sustainability. The same type of argument has been applied to the issue of food miles. For Born and Purcell (2006, p203), in some cases it may be more environmentally desirable to transport food, rather than degrading local resources. For example, growing rice in California or Texas, they argue, would produce environmental costs, associated with water pumping and groundwater depletion, which may well exceed the costs of transport to the environment.

As for the social and economic dimensions of sustainability, it has been pointed out that existing inequalities within a community can mean that economic gains produced by local consumption are allocated in a way that actually exacerbates social injustice – an issue that may have repercussions also at wider scales if the community in question is relatively wealthy (Born and Purcell, 2006, p202; see also Allen et al, 2003). Using the example of California's local governance of pesticide drift, DuPuis et al (2006, p242), for instance, show that 'moving an issue to the local level can serve to make social justice problems invisible and thus disempower marginalized people'.[3] In addition, contrary to positive interpretations of social embeddedness, some researchers notice that farmers' markets, which are often seen as an ideal context to promote relationships of trust between local producers and local consumers, at times actually create social exclusion. This is especially true in countries like the US and the UK, where farmers' markets tend to target largely educated, middle-class consumers, leaving struggling producers and non-affluent consumers to 'weigh concerns with income and price against the supposed benefits of direct social ties' (Hinrichs, 2000, p301).

In short, there are three main issues that potentially strain the relationship between local food systems and the social ideals of sustainable development. First, when based on the interests of a narrow or even authoritarian elite, localism can easily lead to an unreflexive politics that fuels an 'undemocratic, unrepresentative and defensive militant particularism' (DuPuis and Goodman, 2005, p362). Second, food re-localization does not provide a competitive advantage to all places. In fact, whereas 'places' that have high levels of territorial and symbolic capital can easily be commodified, places that do not have a culture of 'terroir' are much less likely to benefit from branding (Feagan, 2007, p37). And third, 'local' and 'green' are not necessarily better than 'global' and 'fair' (Morgan, 2007b). The ethics of a food economy cannot really be assessed without a careful consideration of issues such as production methods and social fairness. In simple terms, there is no reason to think that distant Fair Trade producers are less ecologically sustainable or abide by less just social relations than producers at a local farmers' market whom the consumers know personally (Born and Purcell, 2006, p203).

Interestingly, the critique of localization that has recently been developed in the agri-food literature has also emerged within more general debates on sustainable development. Once considered to be one of the main pillars of green ideology, decentralization has recently been questioned as a necessary condition to create sustainable societies. Carter summarizes this argument by pointing out that the ideal of decentralization is not always reconcilable with the principle of democracy, which, as noted above, represents a core objective of sustainable development. Echoing criticisms of the 'local trap' approach in the agri-food literature, Carter argues that decentralized societies can in fact be parochial and lack diversity or willingness to act. Furthermore, he concludes, global environmental problems such as climate change have a trans-boundary nature; hence they require coordinated national and international action more than local intervention (Carter, 2007, pp58–60).

With regard to nutrition and human health, it has been noted that the local option may indeed be better for products that are difficult to ship (for example, heirloom tomatoes). However, in other instances large-scale farming operations can afford expensive rapid-shipment methods and quick refrigeration that make their produce qualitatively better than small farmers' produce which has spent hours in a truck (Born and Purcell, 2006, p203).

Finally, at a broader and more theoretical level, several writers emphasize how localism is inextricably tied to globalism. In other words, global and local food systems should not be treated separately – they are mutually constitutive and influence and feed back into each other (Campbell, 2004, p346; DuPuis and Goodman, 2005). In some documented cases, local food systems can only become sustainable if they receive support and protection at wider scales (Sonnino, 2007b). In more general terms, this proves that 'directly oppositional stances cannot be successful when they are only local; they require the power of a broader social movement to prevail' (Allen et al, 2003, p74).

In sum, as Born and Purcell (2006) argue, the 'local trap' literature raises the need to reconsider the contributions of local food systems to sustainable development on four main grounds:

1 The local is not always desirable – as expressed by the notion of 'defensive localism', localization can provide the ecological foundations for reactionary politics that run against the ideals of democracy and equity implicit in the notion of sustainable development.
2 Since the local does not always necessarily deliver the desired outcomes of sustainable development, it should be seen as a strategy, rather than as a goal.
3 Indeed, goals such as ecological sustainability, democracy and equity could be more easily achieved through other scalar options.
4 This is especially true when considering that localism is relationally tied to the global and the national: capital has power and local food systems can easily be appropriated.

Born and Purcell's critique drew heavily on the work of economic and political geographers who have argued that scale, as a medium for (and the outcome of) social struggle, is a neutral concept that has no inherent attributes. While this may hold true in the economic and political spheres, some critics argue that these conclusions should not be uncritically imported into the *ecological* sphere. This is the perspective that informs Nevin Cohen's criticism of Born and Purcell, who are alleged to have made unsupported claims – particularly the claim that 'there is nothing inherent about any scale'. Cohen's argument is predicated on the view that some outcomes can be directly related to the scale of agricultural production, including humane animal husbandry, diversity of production, social control and cultural diversity, for example. While it is possible for small-scale dairy farmers to mistreat their cows, he argues that at the other end of the scale spectrum 'a feedlot will rarely be able to operate treating its livestock humanely'. The core of his critique, which is best conveyed in his own words, is that:

> *Living within the scale appropriate for a given ecosystem, and as close to the consumer as possible, is the only truly sustainable method of food production. All other methods deplete the soil, pollute the ground and contribute to climate change through fossil fuel consumption. While we can argue about what scale is most appropriate for what locations and types of farming, the point is that there most certainly is something inherent about scale and we can – with better data – equate a scalar strategy with environmental outcomes. Scale is not 'a fundamentally relational concept' when it comes to biology and ecology.* (Cohen, 2007)

The agri-food debate has not yet reconciled the tension between the emphasis on the local as the most appropriate scale to design sustainable food systems and the recurrent warnings against the dangers of the local trap. In general, the main problem is that spatially deterministic references tend to neglect the materiality of different food products and their differential socio-environmental impacts. For example, transporting luxury products, such as exotic fruit, which are destined to small segments of affluent consumers, is not quite the same as transporting more basic goods, such as wheat and rice, which feed large numbers of people around the globe. The costs and implications of delocalization in these two cases are quite different, and so are the strength and validity of the re-localization argument.

Moreover, while from a strictly ecological standpoint there is no doubt that locally produced and consumed foods reduce food miles, re-localizing the food chain exclusively on this ground may do little or nothing in the context of present efforts to reduce carbon emissions and tackle climate change. In 2007, for example, *The New York Times* published an article reporting that distribution represents just nine per cent of the total emissions associated with a packet of Walkers' chips. The greatest emissions are related to storing and frying the potatoes. Farmers store potatoes in artificially humidified warehouses, which take energy to run. Because of the way they are stored, the potatoes contain more water and take longer to fry. Outlining a position that is gaining some strength in the debate on local food and sustainability, *The New York Times* article concludes that 'when you count the energy used by harvesting and milking equipment, farm vehicles, feedstock and chemical fertilizer manufacture, hothouses and processing factories, transportation emerges as just one piece of the carbon dioxide jigsaw puzzle' (Murray, 2007).[4]

Environmentally, this raises the need to take into account a food product's energy use throughout its *life cycle*, rather than focusing just on the provenance of the ingredients. In some cases, life-cycle analysis shows that conventional food products are more environmentally sustainable than local products. The same is true with regard to the economic and social sustainability of a food chain. Sourcing distant products may have negative ecological implications, but at times it provides a crucial support to farmers in developing countries who rely upon markets in developed countries for their livelihoods. At the same time, rejecting *in toto* the industrialization of the food system, as some local food activists tend to do, means downplaying important issues of food security. As Murray (2007) puts it, 'feeding the world's 6.6 billion people, more than half of whom now live in cities, is not possible without mass production'.

In short, delivering sustainable development is not simply a matter of choosing between global and local. Sustainability embodies quite different (and often conflicting) economic, environmental and social objectives. Their integration requires a complex and dynamic process of mediation and negotiation. For this reason, in the food system, as in any type of socioeconomic system, sustainable development is unlikely ever to be a spontaneous social product. Rather, it demands steering capacity and goal-directed intervention (Meadowcroft, 2007, pp302–303). Democratic governments have a unique and crucial role to play here. Indeed, the school food revolution we analyse in this book highlights the potential of the public sector to establish innovative connections between the global and the local, between conventional and alternative food systems, and, more generally, between producers and consumers. What we are witnessing through the unfolding of this quiet revolution is the emergence of new kinds of food systems that transcend the simplistic dichotomy between local as sustainability and local as trap. Grounded as they are in a new 'sense of mutual endeavour, or perhaps even commitment, to support a production–consumption space that is more human-centred and related to a sense of morality' (Kirwan, 2006, p310), these food systems signal the rise of a new powerful actor on the agri-food scene: a Green State that is seeking to use the public plate to promote economic development, democracy and environmental integration – the triple bottom-line of sustainable development.

Public Food and the Green State: Towards Sustainable Food Systems

To address the fetishism of the local that has so far pervaded debates on sustainable development, some scholars have recently re-emphasized the role of the state as ecological steward and facilitator of trans-boundary democracy. This concept of 'Green State' is predicated on the fundamental assumption that, in democratic societies, the state is the most legitimate and most powerful institution to assume the role of 'public ecological trustee' (Eckersley, 2004, p12), given its mandate, its regulatory power and, as we will see in the next chapter, the enormous scale of its budget.

Broadly speaking, the concept of Green State rests on four fundamental ideas:

1. The state has the *greatest capacity to discipline investors, producers and consumers* (Eckersley, 2004, p12). In fact, the state can affect product properties through, for instance, performance standards or eco-labelling for products and production processes; it can affect product prices through measures such as green taxes and incentives for cooperative consumption; and it can even affect individual behaviour through information and education (Lundqvist, 2001).
2. The (reformed) state is the most powerful actor in *facilitating and nurturing* the *cultural change* that is necessary to promote sustainable consumption (Carter, 2007, p65). In Eckersley's (2004, p245) words, 'the green public sphere is absolutely crucial in facilitating the broad cultural shift towards an ecological sensitivity, in the same way that the bourgeois public sphere facilitated the shift towards the widespread diffusion of liberal market values'.
3. The state possesses *more resources and more 'steering' capacity* than any non-state actor when it comes to monitoring ecosystem change, creating ecological knowledge and solving ecological conflict (Lundqvist, 2001, p457; Barry and Eckersley, 2005a, pxii).
4. The state is the only legal and political institution capable of offering *systematic resistance to the forces of globalization* and to the social and ecological costs of capitalism through its influence not only on investment and production but also on reproduction, distribution and consumption – 'three areas that are often neglected in existing sustainability strategies' (Barry and Eckersley, 2005b, p260).

The Green State debate will be addressed in Chapter 8, but for the sake of clarity it is worth saying here that, in our view, the Green State is more of an ideal type than an actually existing state; it is, in simple terms, a vision of a state that endeavours to take sustainable development seriously. Since this vision is a process, more than an event, it may be better to speak of 'greening the state', an expression that reflects the fact that, as our standards change, so do the demands that we make of the state. In other words, sustainable development in our view is inherently a moving target – that is, it is more about *becoming* than about *arriving*. As we will discuss in the final chapter of the book, greening can be understood in a narrow/shallow sense when it refers to a state that begins to accept its environmental obligations. This is what most states actually mean when they proclaim their commitment to sustainable development. Alternatively, greening can be understood in a broad/deep sense if the state tries to give equal weight to the social, economic and environmental dimensions of development, which is easy to proclaim but much more difficult to translate into practice (Morgan, 2007c).

In institutional terms, the Green State operates in a multi-level governance system that straddles the international, national and sub-national dimensions of state activity. Generally, green strategies first appeared at the sub-national level, where pioneering local and regional governments committed themselves to the objectives of sustainable development long before their national governments. Hence, the sub-national territorial dimension tends to remain the most contentious one for the Green State.

Local and regional action has its advocates and its detractors. The former tend to emphasize that devolved action is part of a decentralized system of trial and error, which is an enormously important attribute when risk and uncertainty are so pronounced. As some of the case studies in this book will show, devolution 'allows experimentation and provides a context where sub-national units can act to address issues that are not yet "mature" on the national scene' (Meadowcroft, 2007, p307). Although local and regional state strategies can be laboratories of sustainable development, whatever little victories are achieved here will necessarily remain spatially isolated and politically fragile unless they are scaled up to, and validated by, the nation-state.

On the negative side, it is argued that neo-liberal state strategies have used devolution to local and regional governments as a ruse to undermine national standards of welfare provision. In other words, they have devolved responsibility without devolving powers and resources. Although this is a valid criticism, it should not lead us to dismiss local and regional action as a matter of principle or, even worse, to make it a hand-maiden of neo-liberalism. The merits of localism and regionalism can only be assessed in the concrete, that is, in terms of social composition and political purpose. In other words, devolution should be judged as being progressive or regressive on the basis of its capacity to create or enhance the things we construe to be intrinsically significant, like deeper democratic structures, social and spatial solidarity, the integrity of the public realm and sustainable development (Morgan, 2004b).

If the Green State benefits from having a devolved structure internally, it also benefits from being engaged externally, not least because the challenge of sustainable development is essentially a collective action challenge to the international community (Stern, 2006). However, the developed and developing countries of the world are not equally equipped to face this challenge; putting it bluntly, the rich countries of the North have the capacity to mitigate the effects of climate change, while the poor countries of the South do not. In the international arena, then, a Green State would try to ensure that pro-poor positions are agreed in multilateral negotiations – like the Doha trade talks or the post-Kyoto agreement on climate change. As we will argue in Chapter 7, the South wants the North to think less about aid and more about development – the difference between means and ends that is often overlooked by rich donor countries.

The school food revolution is an example of the potential of the Green State to mediate between global pressures and local concerns. As we will discuss in the following chapters, throughout the world public authorities committed to the principles of ecological democracy are using the public plate to design food systems that discipline producers and consumers through both regulation and education; promote a new shared vision of sustainable development; and attract significant financial resources but also deal with the social and environmental effects associated with the investment of those resources. These new types of food systems are based on a broadened and increasingly sophisticated notion of quality that embraces much more than just the provenance of food. Indeed, this is a concept of quality that escapes prevailing theorizations proposed in the agri-food literature (see Sonnino, 2009).

To begin with, unlike some academics, the emerging Green State does not confine quality to 'alternative' food networks embedded in place, tradition and trust (Feagan, 2007, p28). In their efforts to achieve the objectives of sustainable development, public authorities involved in the school food revolution often devise notions of quality that also encompass attributes of the conventional food system, such as cost reduction, convenience, consistency and predictability (Harvey et al, 2004a, p3). As the example of New York City in particular will show, in their attempts to meet the goals of sustainable development school food reformers are even capable of co-opting discourses and strategies utilized by the corporate sector.

What emerges in the school food revolution is a type of quality capable of integrating conventions that have thus far been considered almost as mutually exclusive by many agri-food theorists. In particular, proponents of 'conventions theory' argue that different production networks assemble and combine different quality conventions to create seemingly stable (and often irreconcilable) 'worlds of justification'. In the context of agri-food, four main types of quality conventions have been especially emphasized (Murdoch et al, 2000; Renard, 2003; Morgan et al, 2006):

1 commercial or *market conventions*, which define quality through market laws – or, in simple terms, through the price mechanism;

2 *domestic conventions*, which are largely based on trust, face-to-face relations and attachment to place and traditional methods of production;
3 *industrial conventions*, which evaluate goods and products on the basis of their efficiency and reliability and rest on standards and objectified rules; and
4 *civic conventions*, which respond to a set of collective principles and involve goods and products that have general societal benefits.

The Green State involved in the school food revolution does not necessarily privilege one set of quality conventions at the expenses of others. Quite the contrary – as this book will show, in both small rural areas and large cities, in developed and in developing countries, school food reformers have the capacity to assemble and harmonize very different quality attributes. For example, as we will show in Chapter 4, in Rome the school food revolution emphasizes, at the same time, standardization and localization, market values and social inclusion. It is through the integration of this complex (and, to some extent, contradictory) mix of values and priorities that the emerging Green State in Rome has managed to design a food system that promotes, at the same time, economic development, democracy and environmental integration.

From a conceptual and methodological standpoint, then, the school food revolution is also raising the need to overcome the widespread tendency to think of quality as emerging *either* at the demand side *or* at the supply side of the food chain. So far, research on food quality has in fact fallen into two quite separate categories. On the one hand, there are scholars arguing that quality is socially constructed at the consumer level in response to a crisis of confidence in the supply chain that has led to the introduction of new regulations based on health, food and the environment (Allaire, 2004, p63; see also Mansfield, 2003, p4; Renard, 2005, p419). Quality, in this sense, evokes nature and ecology. As in the case of organic food products, it means safety, nutritiousness and accessibility (Harvey et al, 2004a, p3) and is identified with 'a set of physical characteristics that can be measured and standardized and that have a material effect within systems of production' (Mansfield, 2003, p4). On the other hand, there are scholars who stress that quality emerges at the supply end of the food chain, where small producers attempt to gain a competitive advantage on the market through the development of discourses and strategies that ground their products in specific social and territorial contexts (Ilbery and Kneafsey, 2000; Morris and Young, 2000; Sonnino, 2007b). Quality, in this case, implies and relies upon a broad idea of traceability – 'the local, the knowable, the specialized and the exclusive' (Harvey et al, 2004b, p197). As in the case of EU regionally certified products or of Fair Trade products, the emphasis here is mostly on the links between the qualitative attributes of a food product and its producer or place of production (Marsden, 2004, p135).

In the processes of school food reform analysed in this book, quality is not simply an outcome of consumers' concerns about their own safety or of producers' survival strategies. Given its 'general mandate to promote the public good with [...] clear lines of accountability to the general population' (Meadowcroft, 2007, p308), the democratic Green State aims to reconnect food producers and consumers by *integrating* their different needs, goals and priorities.

The emerging Green State also extends the notion of quality beyond the materiality of food products. Various researchers have pointed out that a real shift from a *consumer* society towards a *conserver* society (Carter, 2007, p359) needs a *cultural* transformation in individual attitudes and behaviours towards more sustainable lifestyles (Seyfang, 2006, p387). It depends, in other words, on the creation of new forms of 'ecological citizenship' (Dobson, 2003) that lead people to think critically about social and environmental interactions, engage practically with collective problems and assume responsibility for conduct in private and public life (Meadowcroft, 2007). In this sense, it is important to underline that the school food revolution is not just about food. It is also about new ways of thinking and feeling about food. In parts of Italy, the UK and the US, the Green State utilizes the public plate to educate civil society about sustainable development. Through a wide range of education initiatives and the adoption of a 'whole school approach', the Green State transfers the multiple meanings of food quality to schoolchildren. In so doing, it creates new generations of knowledgeable consumers, capable of protecting themselves from the 'reductionist rationality of the market place' (DeLind, 2006, p126).

In the following chapters, we will analyse the scope and limits of different kinds of quality-based food systems that the school food revolution is designing around the world. We will focus on their historic roots, on their wider regulatory, legislative and cultural context, and on their strengths and weaknesses. Our approach is predicated on the assumption that, in theory, school food has much to contribute to current efforts to meet the challenges of sustainable development. By definition, it is one of the few public services that specifically targets 'future generations'. If properly designed, planned and monitored, it is also a service that holds enormous potential to deliver multiple health, ecological, social and economic dividends, including reducing the human and financial costs of poor diet, lowering carbon emissions, creating new markets for quality food producers and empowering consumers by building their capacity to eat healthily. Through the school food revolution, the emerging Green State is beginning to reshape the spatial, economic and social relationships between producers and consumers and, perhaps most important, to build a *collective* commitment to the three fundamental principles of sustainable development.

2

Procurement Matters: Reclaiming the Public Plate

Public procurement – the arcane process through which public bodies buy goods and services – could be one of the most powerful expressions of a Green State, given its enormous potential to influence behaviour in the private as well as in the public sector. As it is generally deployed, however, public procurement presents us with a curious paradox. In theory, as we said, it is one of the most powerful instruments that governments have at their disposal; in practice, however, it tends to be a somewhat neglected function. Apart from some notable exceptions, the story of public procurement is a tale of untapped potential, so much so that the *economic* significance of procurement seems strangely out of step with its *political* status.

Although they have enormous economic influence, public procurement managers are not free agents when it comes to purchasing goods and services (a constraint that applies with special force to the food sector, given its implications for human health). In fact, the public procurement of food – in schools, hospitals, care homes, prisons and the like – is governed by a bewildering array of national and international regulations. By and large, these are designed to ensure two things: that the *process* is as open and competitive as possible and that the *product* is safe to eat.

Surprising as it may seem, the quality of food served on the public plate – how it is produced, where it is sourced and, of course, its nutritional value – attracted little or no political attention until recently. Indeed, the very idea that public procurement could be a popular subject of debate seemed as preposterous as the notion that the Common Agricultural Policy (CAP) or the Farm Bill could be of interest to the general public in Europe and the US.

What may have been preposterous in the past, however, is certainly not so today. In fact, what type of food is served on the public plate, and what kind of agriculture is subsidized by the public purse, are questions that are no longer confined to a narrow corporatist dialogue between politicians and agribusiness. On the contrary, these critically important questions are becoming mainstream political issues, with more and more sections of society coming to recognize that public food provisioning, far from being a purely agri-business matter, is inextricably linked not just to public health, but also to economic development, democracy and environmental integration – the three principles of sustainable development we discussed in Chapter 1. It is for this reason that a new politics of the public plate is beginning to emerge, particularly in Europe and the US, a politics in which hitherto separate social movements are converging in their demands for a more sustainable agri-food system. Among other things, this new approach aims to prise open the opaque world of public food procurement so that the power of purchase can be deployed for the benefit of the many, not the few.

To examine these multidimensional issues in more depth, the chapter is structured as follows. In the following section we explore the Byzantine world of public procurement, noting that a combination of low public profile and weak political status has obscured the significance of this sector. Here we want to understand both the scope for and the limits to *creative procurement* – that is, a greener, more sustainable form of procurement that takes social and ecological as well as economic considerations into account.

The next section examines the international regulatory system of public procurement. The main focus here is the international system of rules and regulations that define – and constrain – what public bodies are legally permitted to do when they procure goods and services. These rules and regulations are enshrined in World Trade Organization (WTO) agreements that are designed to foster free trade. In the past, agriculture was treated as an 'exceptional' sector on account of its unique status; therefore, it was exempted from international trade agreements designed to open up national markets to global competition. A new round of trade talks, the Doha Round, is currently under way; it aims to 'normalize' agriculture by treating it as just another industrial sector. One of the conflicts at the heart of the Doha Round is a debate between neo-liberalism, with its demand for *free trade*, and the food sovereignty movement, with its demand for *fair trade*. The outcome of this debate will determine the extent to which national governments, especially in developing countries, are allowed to regulate their domestic food systems.

In this section we also examine the rules and regulations governing public procurement in the EU and the US, which aim to ensure that public contracts are subject to open and transparent competition. Significantly, though, public contracts are no longer awarded on the basis of narrow economic criteria. In a number of recent test cases in the EU, for example, public bodies have won the right to award contracts on the basis of wider, more sustainable criteria, creating important legal precedents for creative procurement.

Finally, we explore the new politics of the public plate in more detail by examining the demands that new social movements are placing on the subsidized agri-food systems in the EU and the US. These colossal subsidy systems are attracting mounting criticism at home and abroad – at home because they privilege industrial agri-business over sustainable agriculture; abroad because they undermine producers in developing countries by exporting products at artificially low (subsidized) prices. These agri-food subsidy systems – underwritten by the CAP in Europe and the Farm Bill in the US – determine what types of food are produced for mainstream markets. More often than not, these turn out to be the polar opposite of what we are urged to eat by official public health campaigns, and this creates a massive disconnect at the heart of American and European public policy.

This section concludes by examining one of the most encouraging aspects of the new politics of food – namely, the growing trend for governments to set tougher nutritional standards for public food provisioning in schools. More robust health-promoting standards have helped public procurement managers to raise the quality bar in their own specifications, underlining the point that judicious regulatory reform can fashion new markets in the public sector for quality-conscious food producers.

This chapter shows that the world of public procurement, far from being a law unto itself, is actually embedded in, and subject to, a hierarchy of regulations: from *international* WTO regulations that seek to liberalize the trade in food to *national* regulations that govern the nutritional quality of food. Ultimately, this is the complex context in which school caterers and public officials have to manoeuvre when making decisions about the kind of food children should eat at school.

The Byzantine World of Public Procurement

The public sector is a much bigger economic actor than many people seem to realize. This is especially true in Europe, where the value of the public procurement market reached €1500 billion in 2002, which is equivalent to 16 per cent of the EU's gross domestic product. Straddling government at national and sub-national levels, and embracing a wide spectrum of activities, the public sector constitutes an enormous market for a bewildering array of goods and services, ranging from simple items like pens and paper to complex projects in the civil engineering, military and information technology sectors.[1]

Given their economic and business implications, public contracts are increasingly contested by large private sector suppliers from within and outside the domestic market. Where the public sector client knows what it wants, and has the skills to secure it, the procurement process can be a positive sum game for buyer and supplier alike. However, public sector bodies are sometimes criticized for lacking the business acumen to act as intelligent customers, a situation which makes them overly dependent on the knowledge and skills of their private sector suppliers. One of the most common problems here is that close buyer–supplier links can of course degenerate over time, spawning bribery and corruption.

The Corruption Perception Index (CPI)[2] assessed countries to gauge the level of perceived corruption, which was defined as 'the abuse of public office for private gain'. The most common forms of corruption included kickbacks from public procurement contracts, the bribery of public officials and the embezzlement of public funds. Whatever its shortcomings, the CPI helps to expose the myth that corruption is confined to poor countries, which is clearly not the case given that the US came 20th and Italy was as low as 45th in a league table of 163 countries (see Table 2.1).

Table 2.1 *The 2006 Corruption Perception Index*

Rank	Country	CPI Score	Surveys Used	Confidence Range
=1	Finland	9.6	7	9.4–9.7
	Iceland	9.6	6	9.5–9.7
	New Zealand	9.6	7	9.4–9.6
4	Denmark	9.5	7	9.4–9.6
5	Singapore	9.4	9	9.2–9.5
6	Sweden	9.2	7	9.0–9.3
7	Switzerland	9.1	7	8.9–9.2
8	Norway	8.8	7	8.4–9.1
=9	Australia	8.7	8	8.3–9.0
	Netherlands	8.7	7	8.3–9.0
=11	Austria	8.6	7	8.2–8.9
	Luxembourg	8.6	6	8.1–9.0
	United Kingdom	8.6	7	8.2–8.9
14	Canada	8.5	7	8.0–8.9
15	Hong Kong	8.3	9	7.7–8.8
16	Germany	8.0	7	7.8–8.4
17	Japan	7.6	9	7.0–8.1
18	France	7.4	7	6.7–7.8
18	Ireland	7.4	7	6.7–7.9
=20	Belgium	7.3	7	6.6–7.9
	Chile	7.3	7	6.6–7.6
	USA	7.3	8	6.6–7.8
45	Italy	4.9	7	4.4–5.4
=151	Belarus	2.1	4	1.9–2.2
	Cambodia	2.1	6	1.9–2.4
	Côte d'Ivoire	2.1	4	2.0–2.2
	Equatorial Guinea	2.1	3	1.7–2.2
	Uzbekistan	2.1	5	1.8–2.2
=156	Bangladesh	2.0	6	1.7–2.2
	Chad	2.0	6	1.8–2.3
	Congo, Democratic Republic	2.0	4	1.8–2.2
	Sudan	2.0	4	1.8–2.2
=160	Guinea	1.9	3	1.7–2.1
	Iraq	1.9	3	1.6–2.1
	Myanmar	1.9	3	1.8–2.3
163	Haiti	1.8	3	1.7–1.8

A high CPI score indicates a low level of corruption (and vice versa).
Source: www.transparency.org/news_room/in_focus/2006/cpi_2006__1/cpi_table

Where governments have shown a genuine interest in public procurement, it is invariably because they wanted to use it either to spearhead the growth of advanced technology sectors that were deemed to be of strategic value or to enable domestic firms to act as 'national champions' in international competition.[3] Such policies were pursued with great vigour during the second half of the 20th century by governments across the political spectrum, particularly in France and the US. No country has used public procurement to promote its high-technology sectors more assiduously than the latter, where the procurement policies of the Department of Defense enormously helped US firms to establish leading positions in the software, semiconductor and computer sectors (Morgan and Sayer, 1988).

In Europe, no country has employed public procurement more aggressively than France, where public contracts have been deployed to great effect to modernize key sectors of the economy, such as mass transit, energy and telecommunications, which were tailor-made for the state-led model of French industrial policy (Cawson et al, 1990). On the other side of the European spectrum, the history of public procurement in the UK is littered with costly and embarrassing delays, especially in the defence, information technology and civil engineering sectors, where the biggest and most expensive projects are located. Since the Byzantine world of public procurement is readily apparent here, the rest of this section draws on the trials and tribulations of the British experience to illustrate the nature of the problems.

The fallibility of the British Government as a customer is nowhere more evident than in the defence sector, where the Ministry of Defence is nominally in charge of the procurement process. The scale of the problems in this high-technology sector is without precedent in the UK, even if smaller fiascos like the Millennium Dome in London are more memorable in the public mind. In a recent progress report it was disclosed that the largest 20 weapons projects are currently overspent by almost £3 billion and, taken together, they have been delayed by a total of 36 years. Indeed, such are the cost overruns on the notorious Eurofighter aircraft that the current cost is no longer even published because it is deemed to be 'commercially sensitive' (National Audit Office, 2006).

There is no easy explanation for this lamentable procurement performance, though each of the following factors probably played a part:

- the lack of project management skills at the highest levels of the civil service;
- a bureaucratic culture that extolled policy design over project delivery;
- the silo-based structure at the centre of government in Whitehall, which stymied the dissemination of good practice; and
- the fact that this lack of technical competence is both cause and

consequence of a lack of political confidence, rendering civil servants and their masters reluctant to assert public sector priorities over private sector interests (Cawson et al, 1990; Craig, 2006; Page, 2006).

If military procurement is the worst offender, the UK's record in civil procurement also leaves much to be desired. In 1999, a review of civil procurement in central government exposed a woefully inadequate, if not shocking, picture:

- No one really knew how much the government was spending on a whole range of products and services.
- The government was not utilizing effectively (for example, by leveraging its relationship with suppliers) its position in the market place.
- The fragmented approach to procurement resulted in enormous variations in performance.
- Public procurement was not regarded as a core competence and, as a result, its professional status within government suffered.
- There was plenty of scope for government to become a more intelligent and professional customer, but this potential was not being tapped.
- There were major 'value for money' improvements to be gained simply by doing things better (Gershon, 1999).

These findings have triggered a new era in the history of public procurement in the UK in 2000, when the Office of Government Commerce (OGC) was formed to modernize public purchasing and secure better value for money from government contracts. The modernization programme that followed this catalytic review raised a question that has never been fully resolved and that is relevant to public procurement managers everywhere: should procurement be modernized within an old, cost-cutting business model or should modernization embrace a new, more sustainable, value-adding one? According to Peter Gershon, the architect of the UK's new purchasing strategy, the modernization and greening of public procurement go hand in hand, as he told a Greening Government Procurement conference:

> *Our attention is firmly focused on value for money – not simply the lowest price. This means looking at quality and whole-life costs, including disposal and packaging, which are areas where environmentally friendly products tend to score well. [...] Your task is to work out how to procure environmentally friendly goods while retaining value for money. We should not accept a 'green premium' as an inevitable consequence of greening government procurement* (Gershon, 2001).

Underlying this argument is the notion that the 'green line' is synonymous with the 'bottom line', a notion that procurement managers have a hard job putting into practice, despite its appeal as a principle. Nevertheless, the shock of the Gershon Review was a genuine tipping point, signalling the moment when the strategic potential of procurement begun to be recognized at the highest levels of government. Significantly, this epiphany had occurred in the private sector at least a decade earlier, when firms in the auto and electronics sectors finally understood that strategic sourcing (that is, procurement strategy) was one of the 'secrets' behind the success of Toyota and Nissan (Cooke and Morgan, 1998).

Tony Blair's Government belatedly discovered the value of public procurement around the same time as it woke up to the challenge of sustainable development, a coincidence that led to a spate of new policies for 'greening the realm' (Morgan, 2007b). As part of its new green procurement strategy, the government identified a small number of products – for example paper, timber, electrical products and food – where some 'early wins' could be secured. Of these products, public sector food purchasing received the most immediate attention, largely because of the unexpected political salience of *school food*, an issue that we will address in more detail in Chapter 5.

Long before school food became a *cause célèbre* in the UK, the Blair Government had begun to explore the potential of food procurement in public sector catering. In the wake of the British foot and mouth crisis in 2002, an official inquiry concluded that 'local food' offered untapped opportunities for hard-pressed primary producers to reconnect with their consumers, and it identified public procurement as one of the ways to effect this change (Policy Commission, 2002). This was the political context in which the Public Sector Food Procurement Initiative (PSFPI) was launched in 2003 – a seminal event in the history of public food policy in the UK.

The main aim of the PSFPI was to encourage public sector managers to work in concert with farmers, growers and suppliers to ensure that public canteens purchase from sustainable food chains. Underlying this aim were five broad objectives:

1. to raise production and process standards;
2. to increase tenders from small and local producers;
3. to increase consumption of nutritious food;
4. to reduce adverse environmental impacts of production and supply; and
5. to increase the capacity of small and local suppliers to meet more exacting demand standards (Defra, 2003).

Despite its modest resources, the PSFPI is one of the most innovative programmes of its kind in the world, embracing as it does almost every stakeholder in the food chain, including central and local government, public sector purchasing bodies, primary producers, food service companies and non-governmental organizations (NGOs). Like all such programmes, however, the PSFPI raises the question as to what constitutes a 'sustainable food chain'. One of the features of this kind of food chain is the internalization of the costs that are externalized in conventional food chains by, for example, factoring into the equation the effects on human health and the environment of the entire agri-food cycle. In fact, this is what the PSFPI aspires to do when it speaks of the multidimensional nature of public sector food procurement:

> *If we are what we eat, then public sector food purchasers help shape the lives of millions of people. In hospitals, schools, prisons and canteens around the country, good food helps maintain good health, promote healing rates and improve concentration and behaviour. But sustainable food procurement isn't just about better nutrition. It's about where the food comes from, how it's produced and transported, and where it ends up. It's about food quality, safety and choice. Most of all, it's about defining best value in its broadest sense.* (Defra, 2003)

For the worlds of policy and practice, this statement succinctly captures the multiple meanings of the 'sustainable food chain' we discussed in Chapter 1. At the same time, however, it also uncovers one of the key problems in bringing it into being: the problem of reaching a commonly agreed definition of 'best value'.

Procurement bodies in every country try to rationalize their decisions by claiming to have delivered 'best value' or 'value for money', even though these are notoriously difficult to prove. Perhaps to compensate for poor performance in the past, UK governments now seem intent on trying to transform public procurement into something approaching an exact science, which is an impossible mission, given the many human imponderables involved in the process. Nevertheless, driven by a desire to become a world leader in sustainable procurement, the UK Government has fashioned a curious bureaucratic architecture, along with new rules and procedures, to manage the public procurement process.

At the centre of this Byzantine system sits the OGC, which was given new powers and responsibilities in 2007 to enable it to transform central government's procurement capacity so as to secure more value for money. This latest efficiency drive spawned a new and highly prescriptive procurement process that has no fewer than 15 separate stages to it (see Figure 2.1), each with its own specific guidance.

This new OGC guidance is additional to, rather than a substitute for, all

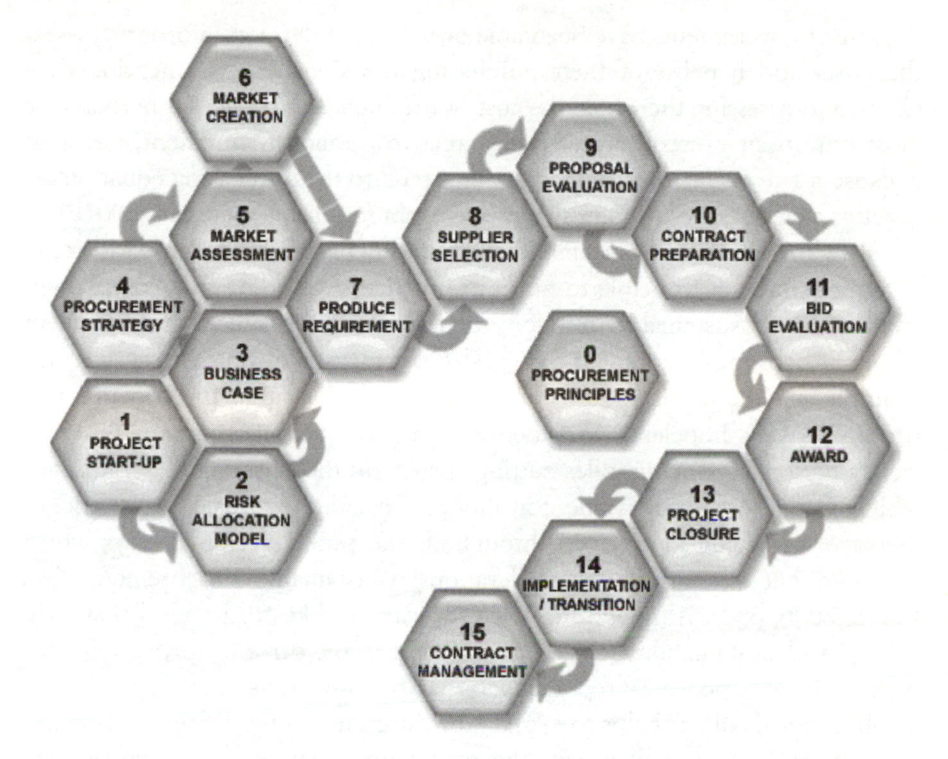

Figure 2.1 *OGC guidance for managing the procurement process*

Source: www.ogc.gov.uk/introduction_to_procurement.asp

the previous guidance issued to public bodies, highlighting the fact that public procurement is becoming an ever more regulated process. The OGC insists that all 15 stages of the procurement process continue to be subject to three overriding principles:

1. All procurement should be subject to competition.
2. All procurement should be conducted on a value-for-money basis.
3. All procurement should be fair, open and transparent, in line with EU regulations.

Since this new wave of regulations might obscure the essential message, the government felt the need to spell it out more explicitly: 'Good procurement means getting value for money – that is, buying a product that is fit for purpose and taking account of the whole-life cost' (HM Treasury, 2007, p4).

Getting value for money, as we have noted, is easier to proclaim than to prove. Indeed, value for money – the modern mantra of governments across the globe – requires a far more sophisticated system of metrics and accounting than most governments have been able or willing to develop in order to assess the costs and benefits of their purchasing decisions. As we mentioned in Chapter 1, assessing the whole-life cost of products and services is perhaps the most important prerequisite of more creative public procurement, not least because it tries to factor sustainability criteria into the purchasing equation – a practice that helps to distinguish *low cost* from *best value* (Morgan, 2007b).

If whole-life costing seems the embodiment of good sense, it is an extremely difficult principle to put into practice. According to a comprehensive study of sustainable public procurement, the failure to implement whole-life costing is one of the biggest barriers to confront. In the UK, for example, this failure is largely attributable to the fact that the procurement profession feels hopelessly torn between two conflicting messages: the 'efficiency message', which implies cutting costs, and the 'sustainability message', which implies securing value for money. In other words, 'the efficiency message was being interpreted throughout the public sector in ways which drowned out sustainability considerations' (Sustainable Procurement Task Force, 2006, p52). Although the UK Government likes to pretend that 'efficiency' and 'sustainability' enjoy equal weight in the drive to secure value for money, the experience we report in later chapters suggests otherwise.

If national culture helps to explain the difference in public sector purchasing patterns, as we will see in the case study chapters, all countries are nevertheless obliged to respect the international 'rules of the game' that regulate public procurement. In the next section, we will try to describe the complex and somewhat ambiguous context that sets these rules of the game for public procurement officers around the world.

Regulating Procurement: The International Rules of the Game

Public bodies are legally required to respect the procurement provisions of international agreements that their governments have signed as members of the WTO, the highest authority on global trade matters. As the biggest purchasers of goods and services in their countries, governments and their public sector agencies together account for a significant share of economic activity, usually ranging from 10 to 16 per cent of GDP in developed countries to as much as 20 per cent of GDP in developing ones. Political pressure to privilege domestic suppliers over foreign competitors can result in closure, where these public sector markets become effectively closed to international competition – the antithesis of the WTO vision of unfettered markets.

In a minority of cases, however, there is more limited agreement, which is formalized into *plurilateral* agreements that are only binding on the WTO members who choose to sign them. The current Agreement on Government Procurement (GPA),[4] which is one of these plurilateral agreements, took legal effect on 1 January 1996, replacing the first GPA, which came into force in 1981.

The primary purpose of the GPA is to open up as much public procurement business as possible to international competition. It is designed to make laws, regulations, procedures and practices regarding public procurement more transparent and to ensure that they do not protect domestic products or suppliers or discriminate against foreign products or suppliers. The current GPA was agreed under the Uruguay Round of trade liberalization talks, which began in 1986 and concluded in 1994. These negotiations secured a tenfold expansion of coverage compared to the 1981 GPA, extending international competition to national and local government entities, whose collective purchases are worth several hundred billion dollars each year. Coverage was also extended to services, procurement at the sub-central level (states, provinces, departments and prefectures, for example) and procurement by public utilities. The new GPA also reinforced the rules that guarantee fair and non-discriminatory conditions of international competition. For example, governments are required to put into place domestic procedures by which aggrieved private bidders can challenge public procurement decisions and obtain redress if they can show that such decisions failed to conform to the rules of the agreement (WTO, 2007). To render the process as competitive as possible, the GPA tries to ensure that neither the location of production nor the nationality of the supplier are deemed relevant factors in public procurement decision-making.

These rules may be part of the conventional wisdom in developed countries, but they are profoundly contentious in many developing countries. In the latest round of WTO trade negotiations (the Doha Round), developed countries are trying to persuade developing countries to open their public sector markets to international competition on the grounds that a 'level playing field' will benefit everyone. Poor countries are not impressed with this free-trade-is-good-for-you argument, which they see as a flag of convenience for developed country firms to gain access to hitherto protected public markets in the developing world. The inability to agree on new rules for public procurement, compounded by the failure to reach an agreement on trade in agriculture, help to explain why the Doha Round has been the most controversial in the history of trade liberalization.

The main reason why the WTO arouses such strong suspicions around the world is because it has expanded the 'reach' of trade policy from arcane issues like tariffs to a range of issues that used to be considered the proper domain of national governments. In other words, across a wide spectrum of sensitive subjects – food safety and standardization, intellectual property rights, services, public procurement and the environment, for example – the WTO has assumed authority in areas that were previously deemed to be *domestic* issues (Morgan et al, 2006).

Mounting disquiet about the 'reach' of the WTO is no longer confined to the developing world. On issues like agriculture, food and public procurement in particular, social movements in developed countries are beginning to mobilize popular opinion against the WTO vision of unregulated free markets, which implies that 'countries would no longer have any control over public purchase' (Hines, 2000, p59).

The WTO vision is also under threat from within, so to speak – that is, from the governments of developed countries that constitute its core membership. Although traditional trade barriers (like quotas and tariffs) have been reduced through successive rounds of liberalization, a series of non-tariff barriers has emerged (like product standards), and these tend to be most prevalent in the agri-food sector. The growth of non-tariff barriers to international trade is now the subject of a highly contested debate in WTO circles, with proponents citing new product standards as a judicious way to reduce consumer exposure to new health and safety risks, while free market critics claim that such standards amount to a new form of 'green protectionism' in developed countries.

Some of the most contentious WTO disputes, such as the EU ban on meat containing growth hormones, Australia's barriers on imports of salmon and Japan's restriction on imports of apples, have involved non-tariff barriers in the agri-food sector. According to the World Bank, non-tariff barriers of all kinds now account for 70 per cent of all global barriers to trade. Such barriers are twice as high for agriculture as for manufacturing, and the highest are imposed by rich countries against middle-income developing countries. Moreover, this is far from being a purely North–South issue: the US complains that it cannot get its top agricultural exports – beef, pork, chicken, soy and corn – into the EU market due to European rules on genetically modified organisms and hygiene regulations. The EU counters by saying that some of the licensing and import practices in the US are designed to discriminate against foreign farmers. As the global trade in processed and perishable food grows faster than trade in traditional commodities, it seems certain that 'standards rather than tariffs will be the greater barrier to such goods' unimpeded journey around the world economy' (Beattie, 2007a).

Clearly, the practice of international trade falls far short of the WTO vision of free and unregulated trade, largely because governments are under pressure to introduce regulations to ensure that products are safe to use, or at least perceived to be so. Here, the precautionary principle is sometimes invoked to justify regulations where there is a high degree of uncertainty as to the effects of a product, under the assumption that the absence of evidence does not amount to absence of harm.

WTO rules are predicated on the neo-liberal idea that governments have little or no role in trade, other than upholding the rules that guarantee fair and non-discriminatory international competition. The GPA was such an important breakthrough for the WTO because, quite apart from its enormous commercial potential, the public procurement market was the most significant example of government involvement in business, and anything that constrains such 'interference' is a good thing in the neo-liberal scenario. The fact that the GPA is a plurilateral agreement, however, means that only a tiny minority of WTO members subscribe to the belief that public procurement decisions should be decided on purely competitive criteria, or that narrow commercial values should be extolled above all other.

But what kind of values should inform the public procurement process? This question lies at the heart of a new debate about green and sustainable public sector purchasing, a debate that has been led by the EU, or more precisely by the 'Green 7' group of countries within the EU, namely Austria, Denmark, Finland, Germany, The Netherlands, Sweden and the UK (Bouwer et al, 2006).

With its 27 member states, the EU constitutes the largest public procurement market in the world. One of the economic pillars of the EU is the Single Market, which is predicated on the free movement of capital, goods, services and people. A legal entity since 1992, the Single Market is still far from being an economic reality, however; in fact, across a swathe of different sectors, particularly the 'public utility' sectors, open competition has been constrained by a range of factors, one of which is the fact that public procurement tends to be biased towards domestic suppliers.

To ensure that public procurement is compatible with the principles of the Single Market, the European competition authorities have devised an elaborate set of regulations to police every stage of the process. Over the past few years, the EU has tried to reduce the Byzantine regulations governing public procurement from four directives to two legal instruments: Directive 2004/18/EC for public works contracts, public supply contracts and public service contracts and Directive 2004/17/EC for the so-called special sectors of water, energy, transport and postal services – in other words, the public utility sectors which are or were in public ownership. Given the enormous confusion that surrounds these directives, especially as to what contracting authorities are actually allowed to do, it is worth quoting in full Recital 46, one of the key parts of the guidance on the use of award criteria:

> *Contracts should be awarded on the basis of objective criteria which ensure compliance with the principles of transparency, non-discrimination and equal treatment and which guarantee that tenders are assessed in conditions of effective competition. As a result, it is*

> *appropriate to allow the application of two award criteria only: 'the lowest price' and 'the most economically advantageous tender'.*
>
> *To ensure compliance with the principle of equal treatment in the award of contracts, it is appropriate to lay down an obligation – established by case law – to ensure the necessary transparency to enable all tenderers to be reasonably informed of the criteria and arrangements which will be applied to identify the most economically advantageous tender. It is therefore the responsibility of contracting authorities to indicate the criteria for the award of a contract and the relative weighting given to each of those criteria in sufficient time for tenderers to be aware of them when preparing their tenders. Contracting authorities may derogate from indicating the weighting of the criteria for the award in duly justified cases for which they must be able to give reasons, or where the weighting cannot be established in advance, in particular on account of the complexity of the contract. In such cases, they must indicate the descending order of importance of the criteria.* (European Commission, 2004)

Short as it is, this extraordinarily dense and carefully crafted piece of legal guidance summarizes the new rules of the game governing public procurement in the EU. Three points need to be made in this regard.

First and foremost, the guidance makes it explicitly clear that contract award criteria must comply with the pro-competitive principles of the Single Market – namely, transparency, non-discrimination and equal treatment of all tenderers.

Second, contracting authorities are not obliged to choose 'the lowest price' tender. Rather, they have the option of awarding the contract to 'the most economically advantageous tender', in which case they have to use criteria linked to the subject matter of the contract in question (such as quality, price, technical merit, aesthetic and functional features, after-sales service, delivery date and completion date), all of which have to be weighted in descending order of importance. All member states must comply with the Directive and allow for the integration of environmental considerations into their public procurement procedures. However, there is an element of discretion left to the member states, which means that 'these are allowed to go further and can, for example, make green purchasing mandatory for certain parts of their administration or set targets' (Day, 2005, p204). This explains the national differences in the implementation of green policies mentioned above.

Third, it must be emphasized that reference to case law has driven the new provisions on award criteria, especially the use of social and environmental criteria. In the famous 'Helsinki Bus Case', the European Court of Justice created a new legal precedent when it ruled that the City of Helsinki was justified in using an environmental criterion – the amount of pollution created by different types of bus – as the basis for awarding a contract for a new generation of buses in the city. It is not overstating the point to say that this case created a new legal basis for green and sustainable procurement in the EU (Morgan and Morley, 2002).

Apart from innovative case law, the rules of the game were also radically changed by the Gothenburg European Council in 2001, when the EU formally committed itself to sustainable development. This created a wholly new context for the evolution of EU policies, especially for public procurement managers, who were officially required to broaden their value sets to include social and environmental factors as well as the traditional economic factor when assessing tenders for goods and services. Following the historic Gothenburg Commitment, the EU was obliged to pursue *sustainable* and not merely *green* public procurement, a distinction that was clarified in the following way:

- *Green public procurement* means that contracting authorities take into account environmental elements when procuring goods, services and works at all stages of the process and within the entire life cycle of procured goods.
- *Sustainable public procurement* means that contracting authorities take into account all three pillars of sustainable development when procuring goods, services and works at all stages of the process.

Clarifying the practical differences between these two forms of procurement, the executive arm of the EU, the European Commission, says that the application of the *environmental* aspects is easier to demonstrate in practice. Green requirements can be specified in the technical demands for the production technology and the selection of materials. Performance and quality standards included in the technical specification can be easily defined and introduced at any stage of the procurement process. In most cases, environmental requirements related to the production process or the product itself can be used to characterize a product and can hence be utilized to describe it in the tender documents of a public tender. By contrast, the Commission states, the specification of *social and ethical* aspects of sustainable procurement is much more difficult to demonstrate, as it is harder to prove their effect on the final product. Additional problems arise in terms of objective verification and quantitative benchmarking of effects and benefits that would allow for accurate and fair evaluation of tenders (European Commission, 2007).

Although this section is primarily concerned with the *international* rules that govern public procurement, hence the emphasis on the WTO and the EU, it is important to also mention something about the *national* rules in the US, given the focus of this book. The US Federal Government spends some US$350 billion annually on goods and services. However, the total public procurement budget is much bigger than that as this figure does not include the purchasing of sub-national governments at state and local levels. At the federal level, two sets of regulations govern the procurement process: the Federal Acquisition Regulation (FAR) and the Buy America Act.

The FAR is an enormous compendium of rules and regulations to which suppliers have to adhere if they want to do business with the federal government and its agencies. Since the FAR has to comply with WTO rules, as does the EU, the ethos of its rules and regulations is very similar to those of these other bodies. To secure best value, for example, Sub-Part 6 requires that 'contracting officers shall promote and provide for full and open competition in soliciting offers and awarding government contracts' (General Services Adminstration et al, 2005). Much later, in Sub-Part 23, the FAR outlines the rules of contracting for environmentally preferable products and services – what the EU calls 'green public procurement'. If we look very closely, Sub-Part 23.703(5) invites officers to realize 'life-cycle cost savings', another key component of green and sustainable public procurement. It is worth quoting the precise sections of these rules and regulations because the order in which they appear in the FAR speaks volumes for the priority that is assigned to them by the Federal Government, a point not lost on contracting officers when they specify what 'best value' means in their award criteria. Although these are federal rules and regulations, they are equally applicable to the procurement process at the sub-national level of government. While the FAR contains rules and regulations that are similar to those of the EU, the big difference lies in the *spirit* of the law rather than the law itself – in Europe, the green and sustainable provisions are projected more prominently and assigned a higher priority.

The FAR does not differentiate the US from Europe, but the Buy America Act most certainly does; indeed, it is often cited in WTO negotiations as a restraint on trade. Originally passed in 1933, the Buy America Act specifies that all federal contracts in the US must use domestic materials and products. Its main purpose is threefold: to promote US-made products, to increase US-based jobs and to protect US manufacturing industry. This presumption in favour of domestic materials and products can be relaxed under certain circumstances, for example if it is impracticable, if domestic products are too costly or if the purchase is covered by a FAR clause. Controversially, however, the Buy America Act was excluded from the WTO Agreement on Government Procurement, though the US is obliged to open its public market to the minority of countries that signed the plurilateral agreement on procurement.

If the rules and regulations in these three jurisdictions – the WTO, EU and US – appear to be clear and unambiguous in principle, nothing could be further from the truth in practice. To illustrate the *practical* problems that face contracting officers who wish to pursue more sustainable interpretations of 'best value', we conclude this chapter by examining how, as a result of the unpretentious demand for more local food in schools, a new politics of the public plate is beginning to emerge in Europe and America.

Procurement in Practice: School Food and the Politics of the Public Plate

School food campaigners are perhaps an unlikely group to have added their voice to the public procurement debate. But, along with a diverse coalition of health and ecological interest groups, they are concerned by the striking disjuncture at the heart of farm policies in Europe and America, where the commodities that are most heavily subsidized by the public purse are not the foods that we normally associate with health and wellbeing. Whatever their differences, these campaigners have come to the conclusion that public procurement is a powerful means through which production and consumption can be recalibrated in a more sustainable fashion, so that farm policies foster, rather than frustrate, health and wellbeing policies. As this new coalition is more visible and vocal in the US than in Europe, we shall focus mainly on the American situation, where one of the key issues is the right of contracting authorities to specify *local* food in their school meals contracts.

The industrialization of the agri-food system was pioneered in the US, where food long ago ceased to have any meaningful connection with place or seasonality, leaving America with a well-deserved reputation for being a 'placeless foodscape' (Morgan et al, 2006). Recently, however, there has been a profound backlash against this industrialized system, which has been triggered by a deep and pervasive desire on the part of American consumers to reconnect with local and regional food producers, to regain the sense of fresh, seasonal taste, and, in effect, to re-personalize the food system by 'putting a face on our food' (Hamilton, 2002).

One of the many expressions of this anti-industrial backlash is the growth of Farm-to-School (FTS) programmes, which aim to:

> *connect school food services with local farmers in partnerships that are intended to bring healthier, fresher food to school meals programmes while at the same time supporting local farmers by providing an additional source of income and a relatively secure market.* (Vallianatos et al, 2004, p415)

Locally focused initiatives like FTS programmes can be fragile creations if they are not part of, and energized by, a supportive national network. In the US case, the Community Food Security Coalition (CFSC) has sought to fill the national vacuum since 1997, when it launched its own initiative, Healthy Farms, Healthy Kids, to promote FTS programmes nationwide (Gottlieb, 2001).

The CFSC includes 325 organizational members in 41 US states, 4 Canadian provinces and the District of Columbia. Its membership covers a wide array of interest groups, including social and economic justice, environmental, nutrition, sustainable agriculture, community development, labour, anti-poverty and anti-hunger organizations, among others. The CFSC aims to build sustainable local and regional food systems that ensure access to affordable, nutritious and culturally appropriate food to all people at all times, and it seeks to ground its activities in the principles of justice, democracy and sustainability. Its support for FTS programmes is therefore part of a much wider agenda to reform US farm and food policy (CFSC, 2007a).

According to the CFSC, over 1000 public schools in 38 states are now buying fresh products from local producers for their school meals programmes, up from a handful of schools in 1998. Although there appears to be significant interest in more local food purchasing from both parents and schools, according to the CFSC 'barriers to local procurement keep this number from being much higher' (CFSC, 2007b). One of the main regulatory barriers here lies in the stance of the US Department of Agriculture (USDA), which insists on interpreting the regulations to mean that local food *cannot* be specified as such by contracting authorities. Perhaps to protect its status as the gatekeeper of the procurement regulations, USDA has consistently objected to alternative interpretations, especially the suggestion that Congress, as part of the 2002 Farm Bill, expressed clear support for geographic preferences in school food service programmes. In a terse letter to all state directors of child nutrition programmes, a senior USDA official said:

> *Federal procurement regulations at 7 CFR 3016.60(c) clearly prohibit the use of State or local geographic preferences. All purchases made with non-profit school food service account funds are to be made competitively, consistent with Federal laws and regulations.* (Garnett, 2007)

A radically different interpretation of the federal regulations has been offered by the Harrison Institute for Public Law at Georgetown University, which argues that USDA has manifestly failed to keep abreast of the underlying statutory basis for the rules and regulations that govern local food procurement. The current statutory basis, it claims, is the 2002 Farm Bill that states (in Section 4303) that:

> *The Secretary shall encourage institutions participating in the school lunch programme under this Act and the school breakfast programme established by Section 4 of the Child Nutrition Act of 1966 to purchase, in addition to other food purchases, locally produced foods for school meal programmes, to the maximum extent practicable and appropriate.*

Drawing on the provisions of the 2002 Farm Bill, the Harrison Institute addresses itself to the key question: can US school districts give preference to local food purchases even when using federal dollars? The answer is:

> *Yes. Some of the confusion has to do with the torturous legal language that provides the basis for the purchasing preferences. Local preferences are not 'required' or 'mandated'. Rather, they are 'not disallowed', and even then, the creation of local purchasing preferences is only allowed because they satisfy the criteria under Federal law for an exception to the prohibition on local preferences! The point is that no one is going to create the preference for you as a buyer – it's up to you to create it, and yes, you are allowed to do so.* (HIPL, 2007)

One would be hard pressed to find a better illustration of the curious nature of public procurement in practice, especially when contracting authorities try to implement a novel or innovative interpretation of the arcane rules and regulations. Although it is yet to appear, the 2007 Farm Bill may help to resolve some of this legal confusion, because, in both House and Senate versions, the language clearly allows contracting authorities to use a geographic preference for the procurement of locally produced goods. Even though geographic preference already looks to be legal, FTS campaigners believe that many food service managers will recoil from using it unless they receive a clear mandate from USDA, the federal agency that provides the reimbursement for the school lunch programme (as we will explain in the next chapter).

Longer term, the fate of the FTS programme will depend on whether US farm and food policy can be rendered more sustainable – and that means committing more money in the Farm Bill to local and regional produce and less to the basic commodity groups. As things stand today, however, the opposite is true. Of the US$112 billion spent on commodity subsidies between 1995 and 2004, more than 80 per cent was devoted to the production of just five crops – corn, cotton, wheat, rice and soybeans (Imhoff, 2007). For decades US policy has focused on developing export markets for bulk commodities, neglecting the development of local and regional markets within the US. Critics point to the enormous imbalance in the Farm Bill priorities, where Congress dedicated less than US$40 million in 2006 to innovative marketing schemes for locally grown produce, 'a tiny fraction of the US$20 billion spent some years on commodity programmes and export subsidies' (FFPP, 2007).

Although new voices entered the Farm Bill debate in 2007, a radical reform of US farm and food policy is not likely in the short term, even though extra funds have been won for new causes, such as infrastructure to support FTS programmes, and for 'speciality crop' farmers who produce fruit and vegetables. Seasoned observers claim that, barring a revolt on the floor of the Senate, 'the fundamentalist pile-it-high philosophy that has informed US farm policy since the early 1970s looks set to endure for the next five years' (Beattie, 2007b).

In the longer term, however, more and more people will contest the Farm Bill as they come to realize that, far from being a narrow farming issue, it is really a 'food bill' that shapes the ecology, the land, the diet and the health of the entire nation. Looking beyond the current Farm Bill, therefore, reformers like Michael Pollan feel encouraged by the fact that:

> *A grassroots social movement is gathering around food issues today, and while it is still somewhat inchoate, the manifestations are everywhere: in local efforts to get vending machines out of schools and to improve school lunches; in local campaigns to fight feedlots and to force food companies to better the lives of animals in agriculture; in the spectacular growth of the market for organic food and the revival of local food systems. In great and growing numbers, people are voting with their forks for a different sort of food system.* (Pollan, 2007)

Does Europe face the same problems when it comes to implementing procurement regulations? The short answer is that the barriers to sustainable (especially local) food procurement (see Box 2.1) have been more apparent in some countries than others, which might seem a somewhat odd answer given that all member states are subject to the same EU regulations.

However, as we will see in the case study chapters, the design of local food procurement policies has been much more of a challenge in the UK than in Italy, arguably the opposite ends of the food culture spectrum in Europe. In fact, of all the European countries, the UK is by far the closest to the US in terms of both its food and its procurement culture. As regards its national food culture, the UK is akin to a 'placeless foodscape', because, like the US, the cultural affinity between products and places was lost long ago, in sharp contrast to Italy, which actively promotes and protects the territoriality of its local food culture (Morgan and Sonnino, 2007).

Box 2.1 *The most common barriers to sustainable procurement*

- **Cost:** Perception of increased costs associated with sustainable procurement. Value for money is perceived to be inconsistent with paying a premium to achieve sustainability objectives.
- **Knowledge:** Lack of awareness of the need for conducting procurement more sustainably and the processes required for this.
- **Awareness and information:** Lack of information about the most sustainable options; lack of awareness of products; lack of monitoring of supplies; perceptions of inferior quality.
- **Risk:** Risk-averse buyers prefer to purchase from suppliers with a good track record. Organizations fear criticism from the media and are therefore less keen to take innovative approaches.
- **Legal issues:** Uncertainty as to what can and cannot be done under existing rules on public procurement.
- **Leadership:** A lack of leadership, both organizational and political, leading to a lack of ownership and accountability at all levels.
- **Inertia:** Lack of appetite for change. Lack of personal or organizational incentives to drive change.

Source: National Audit Office (2005).

The UK procurement culture is also closer to that of the US than to that of continental Europe, with cost-based contracting being extolled over all other values. Having been embedded for so long in this cost-based culture, many public procurement managers in the UK have convinced themselves that they cannot procure local food from local producers because this is prohibited by EU regulations. In reality, these regulatory barriers are more apparent than real, because, on careful reading of the EU rules, public contracting bodies are able to practise local sourcing in all but name. Although it is indeed illegal to specify local products that can only be supplied by local producers (a stance that falls foul of the EU principle of non-discrimination), it is possible to specify for produce such qualities as fresh, seasonal, organic and certified, which allows public bodies to secure local food in practice, even though this is not identified as such (Morgan and Morley, 2002; Morgan and Sonnino, 2007). As a commissioner from Directorate-General Internal Market[5] explained:

> *If it is set out in a non-discriminatory way, [...] it is legitimate to say 'we want foodstuff that is no older than'. If that means in practice that it will have to be locally grown, so be it! It remains a*

> *legitimate criterion, but it is not a legitimate criterion if you say that it has to be produced within 10 kilometres of the school.*

Specifying such qualities as fresh and seasonal is second nature to Italian procurement managers, reflecting the territorial values of a food culture that is actively promoted by national, regional and local regulations, as we will see in Chapter 4. This combination of national food culture and political support helps to explain how and why the Italians interpret EU regulations in a way that underlines, rather than undermines, the territorial values of food. As we have explained above, the UK has experienced major problems in designing more creative forms of public procurement, not least because it has been in thrall to a narrow metric of 'efficiency'. This metric renders it difficult to think creatively about more sustainable forms of accounting, like life-cycle costing for example, where values other than low cost can be factored into the public procurement equation.

If the national interplay of politics and culture helps to explain how and why member states have interpreted EU procurement regulations in such radically different ways, national action is not sufficient to raise the status of public procurement in a Europe of 27 member states. The longer-term prospects for local food procurement will depend on more concerted action to reform the CAP, which is to Europe what the Farm Bill is to the US. Although the bulk of its subsidies are geared to a small number of basic commodities – cereals, meat and dairy, for example – the EU has done more than the US to render its agri-food system more sustainable by diverting funds to rural development, environmental quality and animal welfare. Like the US, however, the EU continues to underwrite a farm and food policy that is largely at odds with its health and wellbeing policies, creating a massive political disjuncture at the heart of the public realm.

In conclusion, public procurement could play an important part in the ecological repertoire of the Green State, but only if its potential was more clearly recognized, if it was implemented by public sector managers who had the competence and the confidence to design creative and innovative tenders, and if the rules and regulations (of the WTO, the EU and the US) were to acknowledge that locally produced fresh food, being vital to human health and wellbeing, should no longer be treated as though it were just another tradable product. None of these changes on its own is sufficient to transform the world of public food procurement, but together they could help to bring the Green State a little closer.

3

Fast Food Nation? Reinventing the School Lunch Programme in New York City

Somehow, we Americans are able to look past the slum housing, the polluted air and water, the bad schools, the excessive population growth, and the chronic unemployment of our poor. But the knowledge that human beings, especially little children, are suffering from hunger profoundly disturbs the American conscience. (Senator George McGovern, 1969, cited in Poppendieck, 1998)

In October 2007, *The New York Times* ran an article titled 'Local carrots with a side of red tape'. The article tells the story of a New York farmer attempting to sell his carrots to New York City's schools. There is no doubt, the article states, that local carrots would produce benefits for both producers and consumers: they 'would help farmers who now mostly grow varieties best suited for the frozen foods industry to diversify' and they 'would be fresher, tastier and take less fuel to ship' (Severson, 2007).

As obvious as this argument seems to be, in reality the carrot producer had to fight many battles for nearly two years before city authorities decided to test his product at a half-dozen schools. His initial attempt to use Sugar Snax, a variety of carrots that can be turned into baby carrots, widely grown in California, failed: as he discovered, New York soil does not grow the best Sugar Snax, and grinding them into baby carrots produced a lot of waste. With the help of a food industry consultant, he then decided to make something called Carrot Crunchers: a type of carrot, shaped like a coin, which creates less waste and is better adapted to the local soil conditions. The new product was a hit in the few schools where it was tested; however, the farmer has not yet invested in special coin-cutting equipment. He probably knows that, in a country like the US, 'it's not a question of saying I believe in local products and I'll buy them next week', as the Director of New York City's School Food Office remarked. It is a question of battling 'a bureaucracy that seems tilted away from local food' (Severson, 2007). The roots of this situation lie in a system that historically emerged as an anti-hunger strategy but that has increasingly developed into a commercialized service, fundamentally adverse to the introduction of high-quality food in schools.

From Hunger to Obesity: A Brief History of School Food Programmes in the US

School food programmes have quite a long history in the US. The earliest examples go back to the 19th century, when, especially in big cities, serving meals to students was a strategy used by volunteer groups or by the schools themselves to fight against hunger and malnutrition. In New York City, the Children's Aid Society ran one of the first recorded school food programmes in 1853; in Milwaukee, the Women's School Alliance of Wisconsin began serving lunches to children in needy areas in 1904; in Boston, the Women's Educational and Industrial Union began providing hot meals in 1908.

With the introduction of compulsory school attendance in 1900, a more widespread movement for 'school feeding' emerged in the US (Poppendieck, 2008). Instrumental in this respect was the publication of *Poverty*, a 1904 book by Robert Hunter that brought to light the problems and dilemmas associated with educating hungry children. As Hunter (1904, p217; cited in Poppendieck, 2008) compellingly argued:

> *There must be thousands – very likely sixty or seventy thousand children – in New York City alone who often arrive at school hungry and unfitted to do well the work required. It is utter folly, from the point of view of learning, to have a compulsory school law which compels children, in that weak physical and mental state which results from poverty, to drag themselves to school and to sit at their desks, day in and day out, for several years, learning little or nothing.*

Along with *The Bitter Cry of the Children*, published two years later by John Spargo, Hunter's work 'brought national attention to widely scattered local efforts to provide meals for poor children at school' (Poppendieck, 2008).

Even though in the following decades efforts to provide school lunches multiplied,[1] the Federal Government did not step into the system until the Depression era, when Section 32 of the Agricultural Adjustment Act authorized the use of federal funds for food donations to schools. With the creation of the Works Progress Administration (WPA) in 1935, women in needy areas were assigned to jobs in school canteens, an initiative that improved the organization of school lunches and promoted the standardization of menus and recipes. By 1941, there were some 23,000 schools serving an average of almost 2 million lunches and employing more than 64,000 people (Van Egmond-Pannell, 1985, p11). To a certain extent, menus, recipes and procedures became standardized.

From the US Government's perspective, the main concern in the early days of these school nutrition programmes had to do with food producers, rather than consumers. The Depression era was extremely harsh on American farmers. As Gunderson (2007) explains, much of the agricultural production went begging for a market, surpluses of farm products continued to mount and prices of farm products declined sharply. To remove price-depressing surplus foods from the market, Law 320, approved in 1936, made available to the Secretary of Agriculture an amount of money to be used to encourage the domestic consumption of certain agricultural commodities (usually those in surplus supply) by diverting them from the normal channels of trade and commerce. Needy families and school lunch programmes became 'constructive outlets' for such commodities (Gunderson, 2007); the idea was that needy schoolchildren 'would be using foods at school which would not otherwise be purchased in the market place and farmers would be helped by obtaining an outlet for their products at a reasonable price' (Gunderson, 2007). In 1939, the number of schools receiving these commodities for their lunch programmes had reached 14,075, feeding a total of 892,259 children (Gunderson, 2007).

From the National School Lunch Act to the Child Nutrition Act

During World War II, the school lunch programme in the US entered a crisis. Federal assistance was cut, commodities were no longer available and nor were WPA workers, many of whom were employed to produce the supplies needed for the war (Van Egmond-Pannell, 1985, p12). Inevitably, the number of schools with lunch programmes declined, reaching a low of some 34,000 in 1943. As a response to the crisis, in 1946 the government decided to formalize the school food service with the passing of the National School Lunch Act (NSLA), which authorized the Consumer and Marketing Service[2] to begin the National School Lunch Program (NSLP). The goals of the programme are clearly stated in Section 2 of the NSLA:

> *It is hereby declared to be the policy of Congress, as a measure of national security, to safeguard the health and wellbeing of the nation's children and to encourage the domestic consumption of nutritious agricultural commodities and other food by assisting the States, through grants-in-aid and other means, in providing an adequate supply of foods and other facilities for the establishment, maintenance, operation and expansion of non-profit school lunch programmes.* (cited in Van Egmond-Pannell, 1985, p21)

For the first time, in the immediate aftermath of the war, the US policy on school lunches began to integrate its original focus on farm support with a new emphasis on issues of health and malnutrition. As Dwyer (1995, p173) states, the initiative was a 'felicitous mixture of good intention and commodity politics'. Indeed, on the one hand the NSLA authorized USDA to provide a broad range of commodities to the school nutrition programmes. By reducing surpluses, this initiative propped up the prices farmers received for their products on the open market (Allen and Guthman, 2006, p404). On the other hand, to safeguard children's health, meals were designed to meet specific nutritional requirements (Nelson, 1981, p29).

The NSLA, however, was not without its weaknesses. As Poppendieck (2008) explains, it did not legally define 'reduced charges' and 'inability to pay'. Local schools were expected to determine who would receive such meals, without, however, receiving extra reimbursements for them. At the time, with WPA projects providing free labour and with the donated commodities, many schools could afford to serve free meals to large numbers of students. But once the WPA was dissolved, and schools had to face the costs of preparing and serving the meals, children's payments became the primary source of funds – a fact that discouraged many schools from identifying needy children.

In the following decades, as the number of children participating in school lunches continued to grow (reaching 14 million in 1960), the school lunch programme increasingly targeted needy pupils. In a series of bills, Congress finally established national standards to define eligibility for free and reduced-price meals and required the Federal Government to provide full reimbursement for all such meals served to eligible children, with no 'cap' on the amount of funds that a state could receive (Poppendieck, 2008).

In 1966, the Child Nutrition Act marked the beginning of a new era for American school nutrition programmes. In addition to initiating a pilot breakfast programme, which targeted primarily poor areas and schools where children had to travel long distances, the Act strengthened the lunch programme by authorizing funds for the Non-Food Assistance Program. By providing financial assistance to public and non-profit schools located in poor areas that needed to acquire the equipment necessary to store, transport, prepare and provide food to children (Nelson, 1981, p33), this Act turned school meals into 'an entitlement for low-income children who attended schools that chose to offer the programme' (Poppendieck, 2008).

Neither were American farmers forgotten by Congress. Indeed, as stated in Section 2 of the Act:

> *In recognition of the demonstrated relationship between food and good nutrition and the capacity of children to develop and learn [...], it is hereby declared to be the policy of Congress that these efforts shall be extended, expanded and strengthened [...] as a measure to safeguard the health and wellbeing of the nation's children, and to encourage the domestic consumption of agricultural and other*

> *foods, by assisting States, through grants-in-aid and other means, to meet more effectively the nutritional needs of our children.*

From hunger to obesity: The commercialization of school food in the US

The 1970s were years of rapid and significant expansion for all the programmes, with Congressional legislation providing funding that enabled thousands of inner-city schools to convert rooms into kitchens and buy the equipment necessary for preparing and serving food (Van Egmond-Pannell, 1985, p17). Underlying this federal effort was a growing concern about hunger in the US, a problem that was brought to public attention by Senator McGovern's Committee on Nutrition and Human Needs. As part of its effort to fight against hunger, in 1971 USDA expanded the selection of schools to receive funds for the School Breakfast Program, targeting in particular schools where there was a special need for improving the nutrition and diet of children of working mothers and low-income families (Nelson, 1981, pp33–34). In 1975, the NSLA was amended to broaden the eligibility criteria for reduced-price meals. As a result, the number of children served free or reduced-price lunches almost tripled by 1979 (Van Egmond-Pannell, 1985, p73). At the same time, efforts were made to improve the quality of the meals, 'under the reasoning that if meals were more acceptable to children, less food would be wasted and children would be encouraged to eat a nutritious diet' (Nelson, 1981, pp34–35). For example, with the Nutrition Education and Training Program, introduced in 1977, initiatives were taken to disseminate nutritional information to schoolchildren and to train school food service providers in nutrition and meal planning.

Counterbalancing the proliferation of federal legislation that aimed to support the school food programmes and emphasized issues of social inclusion, however, the 1970s also witnessed the approval of a law destined to change for ever the philosophy and functioning of the American school meal system. PL 92-433, passed in 1972, opened schools to 'competitive food operations', which, as Senator Edward Kennedy stated, began to place new pressures on local school officials 'to permit the sale of food items that will directly compete with the school lunch and breakfast programmes' (cited in Van Egmond-Pannell, 1985, p20). Years of public outrage and legal battles eventually culminated with the introduction of regulations that placed limits on the sale of four categories of food (soda water, water ices, chewing gums and certain candies). However, a lawsuit by the National Soft Drink Association in 1983 led to an overturn of those regulations, with a federal court agreeing that the Secretary of Agriculture had 'overstepped his authority' and that only the school cafeterias were to be considered under his jurisdiction.[3]

The 1980s were years of progressive deterioration of the school food service in the US. A sharp recession triggered long-term trends towards unemployment and steep cutbacks in federal social spending, which made an increasing number of American people food-insecure. As Poppendieck (1998, p3) recalls, in those years kitchens and food pantries experienced ever longer lines at their doors; in New York City, one hundred new emergency food programmes opened their doors in 1983 alone. The federal budget cuts had a dramatic impact on child nutrition programmes. The prices of school meals sharply increased, with predictable effects on pupils' participation; by 1983, the number of schools serving lunches under the NSLP was down to 90,360, representing a decline of 3975 since 1979; between 1980 and 1983, four million children dropped out of the programme (Van Egmond-Pannell, 1985, p1). High unemployment and the economic crisis were also responsible for a sharp increase in the number of children qualifying for free or reduced-price meals. In 1984, 45 per cent of national school lunches were free and 6 per cent were reduced price. With its percentage of 88 per cent free lunches (Van Egmond-Pannell, 1985, p35), New York became one of the American cities most strongly affected by the economic crisis.

As we will see in Chapter 5 with regard to the UK, during the 1980s cutting the costs of school meal services was part of Conservative governments' anti-welfare agenda. Faced with federal cutbacks, high labour costs and lack of skilled labour, many school districts started to use prepared items and convenience foods and to process the USDA-donated commodities – particularly frozen cherries, whole turkeys, cheese and flour. By 1983, states had processing contracts with more than 500 companies (Van Egmond-Pannell, 1985, pp26–27). The consequences of these changes for the nutritional quality of the meals were dramatic. As Allen and Guthman (2006, p404) point out, a 1992 study found that American school lunches greatly exceeded the recommendations for fat, saturated fat and sodium. In simple terms, 'while the original purpose of having school lunch programmes was that children were not getting enough food, now there is concern that they are getting too much of the wrong foods' (Allen and Guthman, 2006, p404). A representative from the New York Coalition for Healthy School Lunches explained:

> *The four top commodity foods in New York State are ground beef, which ends up in hamburgers almost always, chicken, much of which ends up in chicken nuggets, cheese, which could be mozzarella cheese sticks, cheese that goes with ham and cheese on a bagel, cheese macaronis, cheese on a pizza, there's a lot of cheese. […] And the other item is potato products; most of it ends up in deep fries. The food is also very high in sodium. Some of the school meals have more sodium just in the entrée than a child is supposed to have in an entire day.*

As in the UK, federal cutbacks have affected more than the nutritional quality of the food served to children. Lack of funding has often forced schools to buy cheaper 'heat and serve' meals from national distributors, rather than preparing them on the premises (Allen and Guthman, 2006). Combined with the recent crisis in American public education, this situation has led many schools to promote sales of 'competitive' foods by putting in vending machines, contracting with fast food chains and signing 'pouring rights' contracts with soft-drink companies.[4] Since the amount of money a school district receives is dependent on soda sales, these contracts create 'a conflict of interest between health and profit' (Simon, 2006, p222) that has proven very difficult to resolve, given the lobbying power of soft-drink associations and other business groups and the financial crisis caused by the decreasing support for public education perpetrated by the new right-wing anti-welfare agenda (Guthman and DuPuis, 2006, p434).

Fighting junk food: Towards a school food revolution?

Today, the regulatory and cultural context of school meals in the US is changing. In addition to the recent Farm Bill we discussed in Chapter 2, the Child Nutrition and WIC Reauthorization Act of 2004 has increased access to fresh fruit and vegetables by extending a pilot programme in selected elementary and secondary schools and has provided grants and technical assistance to improve access to local foods in schools through Farm-to-School activities. The Act also established that all school districts participating in school food programmes develop 'Wellness Policies' that promote nutrition education programmes and set nutritional guidelines for all foods available in schools (Food and Research Action Center, 2007).

Soft-drink and snack-food companies are also, at least formally, somewhat embracing the emerging rhetoric of healthy eating in schools. It is hard to tell whether this is the result of actual concern or a response to the difficulties that these companies are facing as a consequence of the introduction of nutritional standards for competitive foods in about half of US states. But whatever the reason, this new approach is producing interesting results. In May 2006, the three top soft-drink companies in the US announced their intention to remove sweetened drinks from school cafeterias and vending machines, starting from the summer of 2008. A few months later, five large snack-food producers also agreed to start replacing unhealthy products in vending machines and cafeterias with more nutritious foods (Kanemasu, 2007, p14).

However, in both cases the agreement and guidelines are voluntary. In reality, a large sector of the soda and junk-food industry continues to resort to neo-liberal values of 'freedom' and 'choice' to justify its presence in the schools. Significantly, as we will see, it is exactly by embracing the same ideals that New York City has been revolutionizing its school food service, reinventing fast food as a healthy option and, at the same time, a means to overcome the stigma attached to school food and to re-emphasize the values of social inclusion that once lay at the heart of the American school food service.

The Complex World of American School Food Procurement

American cities like New York, interested in reforming their school meal service, have to manoeuvre within a very complex governance and procurement system, as noted in Chapter 2. As in Europe, the school meal service in the US is part of a multi-level governance system, with regulations and responsibilities subdivided amongst a number of different federal, state and local authorities. At the federal level, school nutrition programmes are administered by USDA's Food and Nutrition Service (FNS), which is responsible for establishing policies, implementing legislation and providing funds to states. Each state receives cash subsidies from the FNS on the basis of the number of pupils participating in the nutrition programmes. New York State also provides a small reimbursement to offset the cost of feeding pupils (see Table 3.1). As a condition for the receipt of these funds, every year the State's Department of Education, which administers the school nutrition programmes and monitors the performance of the School Food Authorities (SFAs), must prepare a plan providing information about programme participation, proposed use of programme funds and outreach needs.[5]

Funding provided on the basis of the number of meals served does not constitute the major part of the subsidies states receive. The biggest part of these is in fact allocated for free or reduced-price meals, as illustrated in Table 3.1.

Table 3.1 *Reimbursement rates for school lunches*

	Federal	State
Free	US$2.40	US$0.65
Reduced	US$2.00	US$0.215
Full Price	US$0.23	US$0.65

Source: Dykshorn (2007)

These subsidies support children whose families' household income is between 130 per cent and 185 per cent of the poverty level, who currently represent as much as 60 per cent of the total of 29 million children who are fed annually through the NSLP, at a cost of US$7.9 billion.

For some activists, this is an inherently unequal system. As a leader of the New York Coalition for Healthy School Lunches pointed out:

> *A child is receiving a free lunch; the school receives US$2.63 in reimbursements from the federal and state governments for that meal. But if the child is paying the full price out of their pocket for that same meal, they could be paying anywhere from US$1.30 to US$2.75; it's an average range. So the money reimbursed for kids receiving free lunches is subsidizing the meals for fully paying kids. Is that fair?*

In addition to cash reimbursements, schools also obtain commodity, or entitlement, foods from USDA. The FNS's Food Distribution Division is responsible for determining the amounts of commodities to be donated every school year and for coordinating their allocation to individual states. Each state receives an amount of commodities valued at a specific cash amount per meal served under the school food programme (Nelson, 1981). Currently, for each meal served schools receive US$16.75 (adjusted annually) worth of commodity foods. Market availability determines the variety of foods that schools can obtain from USDA, but not all foods on the list are available to all schools; since the food is delivered by truck, there must be enough demand from school food service directors in a particular region in order to get a specific product (Hamlin, 2006). Moreover, schools have access to any surplus agricultural stocks, which are known as 'bonus' commodities. The Federal Government pays for the transportation costs to designated warehouses, where the donated commodities are stored until picked up by the SFA or otherwise sent to processing. The SFA can refuse the offered commodities, but states are under no obligation to replace refused commodities.

In addition to foods like meat and cheese, fresh fruit and vegetables are now available using commodity funds. The programme, called DoD Fresh, started in 1994, when USDA began collaborating with the Department of Defense's Produce Business Unit to pilot the procurement of fresh fruit and vegetables for schools in eight states using a portion of the states' commodity entitlement funds.[6] Due to the success of the pilot initiative, DoD Fresh now operates in 43 states, which have been allocated US$50 million per year of commodity entitlement funds to procure fresh fruit and vegetables.[7] In addition, schools are also allowed to use general funds to purchase fresh fruit and vegetables from DoD Fresh (Dykshorn, 2007, pp4–5).

Finally, at the school district level, the school nutrition programmes are administered by the SFAs, which are responsible for food procurement and services, the approval of eligible pupils for free and reduced-price meals, financial record-keeping and monitoring the school food service operations. In some cases, individual schools or a multi-district consortium purchase directly the food served to children. For the most part, however, SFAs choose their foods from the list of products available from national distributors. Large buyers, like New York City for example, can use their purchasing power to persuade distributors, often through the help of specialized brokers, to supply new products, but this option is not available to smaller authorities. In most cases, procurement officers are also constrained in their purchasing choices by a widespread requirement to accept the lowest bid – a requirement that many cities, including New York, have introduced as a response to a series of corruption scandals.

Although it is important to understand the philosophy underlying school meals in the US, this synthetic description of the governance and procurement process regulating school food does not fully account for the working of the system in New York. With 860,000 meals served every day and an annual budget of US$450 million, the New York City Board of Education is not only the largest in-house school meal provider in the US:[8] after the US military, it is also the largest institutional food buyer in the country (see Box 3.1) – a fact that places New York City in the position of dictating the rules of the game more than any other city in the whole of the US.

Box 3.1 *New York City School Food Department*

Number of pupils:	1,101,000
Number of employees:	9000
Annual Budget:	US$450 million
Percentage of free and reduced-price meals:	70
Breakfasts/year:	32,412,000
Lunches/year:	111,532,000
Snacks/year:	9,908,000
Suppers/year:	2,324,000

Source: FoodManagement (2006a)

If you can do it in New York, you can do it anywhere. This old American adage well captures the potential reach and effects of the school food reform under way in New York City. But it does not do justice to the peculiar challenges that political authorities and procurement officers have to meet in order to feed the most populated, ethnically mixed and socioeconomically diverse city in the US. With more than 8 million residents in the city and a total of 20 million in the metropolitan area, New York is also the most densely populated city in the US. For food suppliers having to deliver daily by truck to as many as 1450 different locations, there are significant logistical challenges to confront.

From a political and governance standpoint, New York is also quite unique. The city is divided into five boroughs (Manhattan, Queens, Brooklyn, The Bronx and Staten Island), each coterminous with one of the five counties of New York State. Each borough has a President, elected by direct popular vote, who advises the Mayor on issues relating to the borough but has minimal executive power and no legislative function. Most executive power is exercised by the Mayor, and this makes the government of New York one of the most centralized amongst American cities.[9]

Cutting Costs, Improving Nutrition: The Challenge of School Food Reform in New York City

In New York City, extreme levels of obesity and hunger coexist side by side. Recent data show that almost 58 per cent of adults are overweight or obese and that as many as 24 per cent of children aged 6–11 are obese, against a national average of 15 per cent. The situation is especially bad amongst ethnic minorities: New York's Latino children, for example, suffer an obesity rate of 31 per cent (Thorpe et al, 2004). As Kaufman and Karpati (2007, p2186) explain, poverty amongst Latino families creates patterns of food acquisition and consumption that have 'potentially negative effects on children, including at various times eating less, overeating and excessive expectations around (often unhealthy) food'. But hunger is also a major issue in New York. In describing his newly created position as Food Policy Coordinator, Ben Thomases explained:

> *The drive really came from the social justice angle, [...] with the idea being that there are two interrelated food policy issues that are really important to the city of New York: hunger, or more precisely food insecurity, and obesity. [...] There is a strong sense that poverty, food insecurity and obesity are very much related issues. [...] We are talking about hundreds of thousands of people that do not consistently have confidence that they can put food on the table.*

Differences in food practices and needs are especially evident in the schools. As the Nutrition Coordinator for the service explained:

> *You have some areas in New York where [...] you may get issues like 'Why can't we have hummus? Why can't we have tahinis? Why can't we have soups?' as opposed to a different neighbourhood with a different socioeconomic background where all they want to know is 'Hey, why can't we have seconds?'. [...] Where some*

> *people worry about food security, food safety, others may just want to make sure that the child can eat at a certain time.*

In this kind of context, New York City's school food reform began with a strong focus on the only issue that offers the potential of reconciling different, or even conflicting, food consumption habits, priorities and needs: nutrition. Indeed, when Mayor Bloomberg and Chancellor Joel Klein launched the Children First initiative in 2003 to improve New York's school system through changes in leadership, management, planning and strategy, new nutritional standards for all food sold on school campuses (including cafeterias, school stores and vending machines) were introduced. As David Berkowitz, the Executive Director of the New York City Board of Education Office of School Food and Nutrition Services (also known as SchoolFood), stated, 'We have a social mission, and nutrition is at the core of what we do' (FoodManagement Staff, 2006a).

Improving nutrition was not the only goal of the reform. Equally important in Bloomberg's mind was the need to make the school lunch programme more financially viable. At the time, the SchoolFood office used to pay the invoices accrued by each individual location. This system, which required no accountability for specific schools, had contributed to huge budget deficits – for example, in 2001 a deficit of US$75 million had to be paid through tax subsidies[10] (FoodManagement Staff, 2006a).

The dual goal of reforming the school lunch programme by raising the nutritional quality of school food while, at the same time, streamlining its operations, led to an overhaul of the city's procurement approach. Berkowitz, a professional with a background in food management, was hired in 2003 'to bring some of the best practices from the private sector [...] to the department, while maintaining self-operation' (FoodManagement Staff, 2006a). Shortly after, chef Jorge Collazo was brought on board to create a Culinary Concepts Team, formed by regional chefs (one for each of the five boroughs), which was given the responsibility of raising the nutritional quality of school food.

Under Berkowitz's leadership, a series of initiatives were taken to streamline and simplify the functioning of the school lunch programme. At the onset of the reform, SchoolFood attempted to entrust the service to a single distributor. However, once it became clear that the amount of food used by city schools was too much for one distributor to handle, the City decided to contract four distributors to supply the five boroughs. In compliance with a city procurement law requiring acceptance of the lowest bid, distributors are selected on the basis of the proposed price and are awarded a five-year contract, with the option of renewing it for another two years. Unlike what happens in Rome (see Chapter 4), socio-environmental costs are not taken into account in the evaluation of the bids submitted. As a procurement officer simply put it, 'If you start saying "your trucks have to be environmentally friendly", it's going to affect the bid. You are no longer looking at this as a low bid.'

In addition to changing the distribution system, New York also attempted to make the school lunch programme more accountable, by giving the regional coordinators responsibility for their own 'profit and loss' statements. This created a chain of responsibility and accountability from the managers (who oversee 3–5 schools) to the district supervisors, who report to the regional coordinators. As a result of this initiative, the deficit has been reduced to approximately US$30 million (FoodManagement Staff, 2006a).

To meet the other fundamental goal of improving the nutritional quality of the meals served, Collazo had to face enormous challenges. When he arrived, the schools were used to 'horrible, ghastly products: convenience foods lacking in nutrition and flavour'. Implementing a quality revolution like the one that happened in Rome (see next chapter) was prohibitive in a city where schools, which are on average more than 65 years old, tend to be equipped with modified kitchens that contain just a convection oven and, at times, a steamer, making it impossible to cook the food on premises. As Jorge explained, 'When I came, every piece of chicken and fish was breaded, or fried, or had some kind of coating on it. I asked, "Why not a plain chicken breast or a plain piece of fish that we could put a good, low-fat cacciatore or guisado sauce on, and then reheat?"'. It turned out that neither the plain fish and chicken nor the sauces were available (Spake, 2005).

Commodity foods represented another obstacle to SchoolFood's mission to improve the nutritional quality of the menus. Berkowitz explained:

> *They have turkeys, now we have these turkeys and we have to process them. Now we would prefer not to have the dark meat because it is fattier, but you can't say 'We just want the white meat' – you just have to take the turkey. There are truckloads of turkeys. Whereas if we weren't getting these products, we wouldn't buy dark meat.*

To create items that are nutritious but also easy to prepare, Collazo had to establish close relationships with food manufacturers and distributors,[11] who, as he explained, initially did not seem to be aware that there were healthier products available or that it was possible to procure something more nutritious. However, with an estimated US$123 million annually to spend on food, he had significant power in convincing manufacturers to create new products, such as low-sodium sauces and leaner meat products (Schibsted, 2005, p37). As part of his job, Collazo has spent quite some time working with food manufacturers to help them devising 'precooked protein and other products that can be placed in the oven but which have an appearance "that speaks of freshly made"' (Culinary Institute of America, 2005, p13).

After working on improving the quality of the products available, Collazo had to face another challenge: to create a standardized menu that was flexible enough to accommodate the varying food preferences of New York's population but that was also strict enough to allow quality control. His goal, as he explained, was to 'have a menu day that says "roast chicken selection"; this could be just a roast chicken with some herbs and a vegetable and bread component at one location, while at another it could come with a Dominican tomato-based sauce or with Teriyaki flavouring' (FoodManagement, 2006b).

In a country where cooking and healthy eating skills are on the wane, Collazo and his team needed to work quite closely with the cooks to help them implement the new menus. Most of the work was culinary training – in Collazo's words, 'basic things like how to roast vegetables, how to defrost properly or how to batch-cook vegetables rather than cook them all at nine o'clock for your lunch service time' (FoodManagement Staff, 2006a). At the same time, Collazo and his team also assisted cooks in setting up salad bars and other special programmes, including burrito, deli and pasta bars.

Involving parents was an important aspect of the reform for New York City's authorities. The main tool utilized to achieve this goal is the School Food Partnership, a mechanism that resembles the Italian Canteen Commissions we will describe in Chapter 4. Under the guidance of Herman McKie, the Nutrition Coordinator for SchoolFood, the Partnership provides an arena for parents, students, administrators and various non-profit organizations to discuss questions and concerns about school food and to educate the community at large about nutrition. The School Food Partnership holds monthly meetings throughout the boroughs, distributing posters, brochures and recipes. In addition to educating the larger community about nutrition, the Partnership attempts to take parents on board by making them aware of the improvements introduced by the school food reform. As the Regional Coordinator for Queens explained:

> *Many of the students would go home and if their parents ask, 'What did you eat?', the kids will say, 'Nothing. There is nothing to eat. It's terrible, the same thing everyday.' So what does the parent know? This is what the child is telling them. And we are able to bring them in and say, 'Hey, come to the next partnership, we will show you the menu, we will educate you.'*

The ultimate goal that procurement officers pursue by seeking parents' involvement in the reform is a simple one: increasing students' participation in the school food programme. As explained above, the financial viability of school food programmes in the US is heavily dependent on the number of meals served. In a context where school lunches are traditionally stigmatized as 'welfare' food, boosting participation is an important but very challenging task. As the Regional Coordinator for Brooklyn said:

> *Even among lower economic people there are castes, there is social strata – those who have to accept welfare and those that don't, even though they are all poor. [...] There is always something that is set up as the point of differentiation. And in the case of high schools and junior high schools in this area it is 'Do I have to accept the food? Or can I not accept it?' [...] It is a very hard thing to overcome; it is really engrained in the children. So hard to overcome – and sometimes it just breaks your heart, because you know that they are hungry. And we ask them, 'Why don't you take something?', and they say, 'No, I can't.'*

In New York, the SchoolFood Office has introduced a number of initiatives to improve participation, including:

- *Targeting parents.* To convince parents to fill out the eligibility forms for free and reduced-price school lunches, SchoolFood runs contests that offer prizes such as a holiday in Hawaii. This initiative, however, has not been very successful. In fact, in a city with a high proportion of illegal immigrants, many parents continue to be reluctant to fill out forms (Dykshorn, 2007, p13).
- *Changing the way students pay for school lunches.* Currently, students paying full price for their school lunches use cash, whereas children receiving free or reduced-price lunches pay with tickets or swipe cards. To eliminate the stigmatization created by this difference, SchoolFood is rolling out in the high schools a uniform card swipe and PIN system, to be used by all students, which would also allow parents to pre-pay for school meals, as happens in many other parts of the US.
- *Free breakfast for all.* In 2003, after realizing that it was more cost-effective to provide free breakfast to all students, rather than paying school employees to collect money from the few students who were actually paying for their breakfast, SchoolFood introduced a universal free breakfast for all students. Providing free breakfast is seen by New York authorities as another strategy to remove the stigma attached to school food. However, even though the total number of breakfasts served has increased by 5 million over the last three years, just 10 per cent of New York pupils participate in the programme. As a result, in 2006 SchoolFood missed out on nearly US$134 million in federal funds allocated for breakfasts (McGovern and Quinn, 2006).

Despite the mixed results achieved through these kinds of initiative, pupils' participation in the school lunch programme has been going up in New York City, particularly in high schools, where a three per cent increase in the total number of meals served has been recorded in the last three years. This success is the result of a radical, as well as controversial, change in the cultural context of school food. As the Head of SchoolFood's Marketing Office simply put it:

> *I don't like to call it fast food, because you can also start thinking about unhealthy food. But it's really to give them what corporate America is doing. To mimic what corporate America is achieving.*

Children as Customers: New York City's (Fast) Food Model

New York City's authorities operate in a socio-cultural environment of food choice and procurement that, as noted in Chapter 2, is radically different from the one we will describe for Italy. In the US, food corporations spend as much as US$15 billion a year directly targeting children with junk food marketing and advertising (Birchall, 2007). As a result, it has been calculated that American children spend annually almost US$30 billion on foods that are detrimental to their health (Nestle, 2006).

For the school food service, competing with fast food and junk food is a daily battle. Most middle and high schools allow children to leave at lunchtime, and this provides an opportunity for them to go to one of the many fast food restaurants that are strategically located around the schools. At the same time, budget deficits are leading many schools to ignore the Wellness Policy introduced in 2004 and organize fund-raising activities around the sale of junk food. As the Nutrition Coordinator described it:

> *Even in elementary schools last month they were selling candy right in the back of cafeteria during lunch. [...] They are not supposed to be doing it, but there has to be a balance: 'How do I pay for my science project? My kids want to go to the museum next month. How do I pay for this?'. And they do it by selling candy, they do it by pizza parties and school stores.*

Similarly, a procurement officer explained:

> *What do you tell the principal that says 'if I can't sell this, then I can't get library books, or take kids on a field trip'? I mean, you can't be the nasty bureaucrat that says 'No, you can't sell that!'. The principals are looking for money for their schools, and if Pepsi*

> *comes through or Frito Lay comes through, Subway comes through and says 'listen, we can give you a couple of thousand dollars if you let us in' [...] the principals are looking to do the best they can for their kids and this is where things come in conflict.*

For many activists, this competition is perhaps the biggest barrier to overcome in the reform process. As the leader of the New York Coalition for Healthy School Lunches put it:

> *As long as there's unhealthy food that kids will prefer, then most of them are not gonna make healthy choices. So my feeling is this: I don't care how much healthy foods we've added, until we get rid of the unhealthy foods, we're gonna have a problem. You can have a wonderful school meal. But if they still sell the cookies, the ice cream and the chips at the end of the line, you're gonna have a lot of kids who will buy cookies and chips and ice cream for lunch instead of eating the healthy school meal.*

Like many other cities in the US, the SchoolFood office has implemented a controversial strategy to deal with the problem. Rather than imposing a healthy eating rhetoric that would probably find very few followers amongst New York pupils, SchoolFood has chosen to serve healthy meals *disguised* as fast food. Social marketing campaigns that directly aim to increase pupils' participation in the lunch programmes and merchandizing initiatives focused on enhancing the dining experience have been a fundamental aspect of the reform. Santa DiSclafani was specifically hired to work with chef Jorge Collazo in an effort to create school cafeterias that resemble restaurants and appeal visually to students. Using SchoolFood's slogan 'Feed your Mind' and blue and green logos, SchoolFood is creating a 'brand identity', as DiSclafani stated, while also changing the presentation of food items. For example, paper or plastic wrapping of certain foods, such as breakfast burritos, have been introduced to make the meals resemble what children would be served at a fast food restaurant (Schibsted, 2005). According to Collazo, it has been an effective approach, as demonstrated by children's reaction to the salad bars:

> *I said 'We really need to pump up salad bars in schools' and the reaction was 'Oh no kids don't like green stuff', and I said 'Yes they do, but it's not being presented properly'. Kids are much more sophisticated than they used to be, they eat out, so they like what we like, well presented, clean and good [food]. This is not something you have to get from the Dalai Lama – they want a good, well-presented product.* (Wright, 2007)

Dishes are prepared, named and presented with an eye to what children would look for in places with pricier menus. When deciding on new dishes, Collazo's team always consults pupils for suggestions. As he pointed out, 'Surveying your customers is a critical part of meeting their needs, regardless of what business you are in' (Schibstedt, 2005, p37). In an effort to increase sales of certain products, Collazo has been running successful sweepstakes, which have been responsible, for example, for a 15 per cent increase in the sales of veggie burgers (Schibstedt, 2005). Theme days, such as 'The 100th Day of School', 'Fortune Cookie Day' or 'Caribbean Day', have also been introduced as part of the marketing effort.

Outside the schools, the message has been reinforced by local sports stars, through, for example, public radio announcements and designated days for high school sports teams to wear their jerseys to breakfast and eat as a team, showing other pupils, as Collazo said, that 'like sports, school food is cool'. At the same time, Collazo worked to change the atmosphere of the school cafeterias. Central to this effort was the formation of a partnership with the Institute for Culinary Education, which every year holds a two-day training session for school cooks to teach them to treat pupils as 'customers' and to view the dining hall as a franchise restaurant, rather than as an institutional feeding centre.

Children are responding well to SchoolFood's social marketing campaign. The Regional Coordinator for Queens underscored the positive impact of giving children a choice:

> *When we opened [the deli bar], the kids said, 'Oh, this is like Subway' and the kids were saying, 'Oh my gosh, we have a Subway in our schools' and they loved it. [...] They felt empowered. [...] They want to feel like young adults and have choices. [...] If you label something 'healthy' they think, 'I am not going to let them make me healthy.'*

This is especially important in the most deprived areas of the city. In describing the early results achieved in areas such as Bushwick, the Brooklyn Regional Coordinator said:

> *The kids sort of put their guard down when they see the burrito bar or the mix-it-up salad bar or the deli bar. That excites them. Something there reminds them of the fast food they buy on the outside, even though it's much more nutritional.*

The social marketing and merchandizing campaigns that lie at the core of New York City's school food reform are, for public authorities, a way to fight back against the overwhelming power of corporate America. As the Head of SchoolFood's Marketing Office explained:

> *Kids' attention is captured by the multi-million-dollar ads. What we are trying to do is to capture their attention. And [corporate America] have the money to have advertisements constantly geared at these children. [...] If we don't create a marketing programme that's what they are used to seeing regarding publicity, we are going to lose them. Because [...] they walk right outside of school, and in the same block it's McDonald's, it's Wendy's, it's Taco Bells. So if they are not held inside the school [...] if they are allowed to walk outside, we lose them.*

While enjoying their fast food dining experience, New York schoolchildren are actually getting healthier meals. Thanks to the efforts of Collazo and the Concept Culinary team, the quality of school food has improved in recent years, achieving nutritional standards that exceed those recommended by USDA. Every day, schools offer two main meals in addition to a vegetarian option. Meals now include more fish and foods with plant-based protein; oils and French fries are trans-fat-free; fresh and frozen (rather than canned) fruits and vegetables are offered every day. Artificial colourings, meat mechanically taken from the bone, hormones and artificial sweeteners have all been eliminated from New York's menus (Schibsted, 2005). In addition, ethnic food is widely offered in the schools, salad bars have been placed in all high schools, and, in an effort to curb childhood obesity, whole and flavoured milk have been replaced by low-fat and skimmed chocolate milk in all schools (Garland, 2006).

Unlike Rome, New York City does not have the financial resources in its school food budget to procure organic food. Local sourcing is also very difficult, given the short growing season of fresh produce in the area, the insufficient supply capacity of individual farmers and the limited packing and distribution capacity in the region (Market Ventures et al, 2007, p86). Nevertheless, local products have been actively introduced in New York's schools, making the city's 'fast food model' much more sophisticated and complex than it would at first glance appear to be. Everything started in 2003, when, as part of its Farm-to-School efforts, the New York State Department of Agriculture and Market convinced SchoolFood to promote state farm products. Through the adoption of a creative procurement approach, SchoolFood wrote a specification for fresh apples that identified a variety only grown in the state. During the first year, 5.5 million pounds of local apples were purchased, providing an economic benefit to the state's agricultural economy of almost US$1.5 million (Market Ventures et al, 2007, p85).

One year later, the City decided to use its entire DoD Fresh allocation to source sliced apples that are grown, processed and packaged in New York. As Berkowitz explained, this recent initiative is a good example of the effectiveness of the City's reform strategy:

> *When we sold [...] whole apples, the kids played baseball with them; when you give them a sliced apple, you can't keep up with them.*[12] *It tastes good and looks good. To me, that is the merger of marketing and nutrition.*

In 2005, SchoolFood entered into a partnership with SchoolFood Plus, a US$3 million collaborative initiative, funded by the W. K. Kellogg Foundation, which aims 'to improve the eating habits, health and academic performance of New York City public schoolchildren while strengthening the New York State agricultural economy through the procurement of local, regional produce' (Market Ventures et al, 2005, p17). In addition to devising, implementing and promoting new plant-based recipes and organizing activities such as 'CookShop Classroom' to teach children how to cook, SchoolFood Plus has a local procurement agenda. One of its members, Karp Resources, took on the role of public interest broker to act as a facilitator and deal-maker between local growers and packers and the four distributors contracted by SchoolFood. After making considerable efforts to gain the trust and participation of both distributors and SchoolFood, in 2006 Karp Resources was given the opportunity to source 130,000 pounds of local plums, peaches, nectarines and pears for summer meals, for a total value of almost US$50,000 (Karp Resources, 2007, p2). The same year, SchoolFood contracted a yoghurt producer who uses local milk to supply approximately 7000 cases of yoghurt per month.

Much can still be done to increase the scope for local food in New York. However, the progress the city has been making in this direction should not be underemphasized. In the largest school district in the country, 'one with some of the oldest facilities and highest labour costs as well as one of the most politically visible nutrition programmes' (FoodManagement Staff, 2006a), the school food revolution can only advance by small steps, or, as a representative from SchoolFood Plus put it:

> *with us taking the educational approach, saying 'look what we could do, let's start off small in maybe one or two schools, see what we can do, see the positive benefit'.*

Making Sense of New York City's (Fast) Food Model

For some activists and academics, initiatives like the school food reform implemented in New York City are fundamentally flawed because they are informed by a supposedly neo-liberal philosophy. There is some truth in this argument. In New York, values and practices from the private sector have been consciously brought into the school food reform: from the initial attempt to consolidate the food distribution system to the efforts made to create a 'brand identity' in the cafeterias; from the emphasis on individual choice as a mechanism to trigger change to the widespread view of children as 'customers'.

However, dismissing the efforts made by school food authorities on the basis of the approach utilized, as some tend to do (see, for example, Allen and Guthman, 2006), is, in our view, naïve. Indeed, this kind of criticism fails to take into consideration the fact that, in a country like the US, public authorities operate within a larger political, regulatory, economic and socio-cultural context that is in many ways hostile to the creation of sustainable school food systems. As we have argued, this is a context in which:

- cooking facilities and healthy eating skills have been progressively lost in the last two decades, forcing schools to buy 'heat and serve' meals from national distributors, rather than cooking them on-site;
- the stigmatization of school food as 'welfare food' negatively affects pupils' participation in the school lunch programme, creating significant financial difficulties to schools in a system where school meals are funded through reimbursements;
- school food is bound to compete with different kinds of nutritionally poor foods: from the donated commodities that the Federal Government 'dumps' on the schools to the competitive foods sold 'à la carte' during lunchtime to the fast food immediately available next to the school building to children who are allowed to leave at lunchtime; and
- children's food habits are heavily influenced by the massive marketing campaigns of junk food and soft-drinks companies.

This is not a context that fosters radical action at the local level, as occurred for example in Rome (see Chapter 4). Nor is it a context that encourages local authorities to even consider the links between public food and environmental sustainability, as occurred in London (see Chapter 5). Rather, it is a context in which a sustainable procurement approach requires changes at multiple levels and scales. In simple terms, 'neither the Federal nor State Governments have a great deal of control over local programmes, and yet changes at all of these levels will be needed to change what is served in the nation's schools' (Dwyer, 1995, p176). As a New York procurement officer put it:

> *Let the Federal Government decide that we don't accept any produce from outside the country. [...] If the city sets a regulation that all vehicles must pass the emission tests [...] if the State decided that all produce must be grown locally to get reimbursement. [...] Those are the ones that set the laws. We do self-impose when we can, where it makes financial sense for us and it doesn't affect the scope of the contract.*

In short, it is the socio-cultural environment of food choice and procurement that must change before anything else happens. And this will not happen overnight. As our examples from the UK, in particular, will demonstrate, it may take years to reconstruct a service that has lost its core values to the commercialism of the market. In this respect, New York City must be praised for its attempts to counterbalance the market and industrial conventions that still dominate the school lunch programme with a renewed emphasis on social inclusion. Since the beginning of its school food reform, the city has provided free breakfast to all of its schoolchildren. It has promoted ethnic dishes that reflect and celebrate cultural diversity. It has introduced initiatives, such as the card swipe and PIN system, to fight against the stigmatization of school food. At the same time, New York has embraced an inclusive procurement approach that is building new relationships of trust and knowledge-exchange within and beyond the school food system. Parents, non-profit organizations, State departments, distributors, farmers, cooks and school personnel have all, at various stages, actively contributed to a reform process that has raised the nutritional value of school food, enhanced children's participation in the school lunch programme, reduced budget deficits and created opportunities for local sourcing. A shared vision of equity and democracy around school food is perhaps beginning to re-emerge, as New York City leads America on the thorny path towards sustainability.

4

School Food as Social Justice: The Quality Revolution in Rome

> *It's a paradox that the country of the Mediterranean diet needs to introduce nutritional guidance because of advertising, consumerism and the hurry people are in.* (Italian Health Minister, February 2007, cited in Kington, 2007)

Recent data on Italian food habits depict a schizophrenic country. On the one hand, contrary to prevailing assumptions and widespread stereotypes about Italian healthy food habits, it appears that the Mediterranean diet continues to give way to fast food, with an estimated 600,000 Italians eating at McDonald's every day (Helstosky, 2006, p156). At the same time, processed foods rich in fat, sugar and salt seem to be especially popular amongst children and teenagers. In 2000, 20 per cent of Italians between the ages of 6 and 17 were overweight and 4 per cent were obese (Brescianini et al, 2002). In 2005, overweight and obesity levels among Italian children aged 7–11 exceeded 30 per cent (International Obesity Task Force, 2005).

On the other hand, ethical, informed and sustainable consumption habits are steadily on the rise in Italy. According to a 2005 report, the food choices of 41 per cent of Italian consumers are driven by the desire to avoid GM or contaminated foods (Naselli, 2005). In 2006, another survey showed that 83 per cent of Italians carefully read food labels, 71 per cent make purchasing decisions on the basis of country-of-origin labelling and one Italian out of four habitually consumes organic products. Health and safety concerns are not the only factors behind these trends. The number of 'reflexive' consumers in Italy seems to be steadily growing, with 33 per cent of the people sampled purchasing Fair Trade products and another 15 per cent boycotting specific brands and products for ethical reasons (Ceccarelli, 2006; Diamanti, 2006).

Clearly, like most Western countries, Italy has been experiencing significant changes in food consumption patterns linked to the popularity of convenience foods, the decreasing amount of time devoted to the preparation of meals and the falling share of money devoted to food in household disposable income (Morgan and Sonnino, 2005). However, unlike in most Western countries, *in Italy the public sector has been actively fighting against the generalized deterioration in the population's nutritional habits through political action.*

In highlighting the role of public authorities in shaping popular consumption habits, Helstosky (2006, p151) argues that in Italy 'government intervention was perhaps more important [...] than cultural traditions were'. In this respect, it might be stated that, in contrast with countries like the UK and the US, where sustainable school food systems often develop *despite* government action, in Italy the school food revolution is happening *because of* state action.

Historically, it was in the aftermath of fascism that the Italian national government began to monitor food consumption habits as a measure of democracy's success. At the time, Helstosky (2006, pp129–130) explains, 'social equality and health replaced nationalism and self-sufficiency as key governmental concerns'. As Senator Giuseppe Aliberti wrote in 1952, 'the war started a welfare movement that, in the post-war period, became widespread through the construction of canteens, which represented great progress in terms of food consumption for the masses' (cited in Helstosky, 2006, p132).

In this context, collective feeding began to receive special attention, to the extent that 'nutritional levels became one way to measure whether the Republic delivered its promise of freedom and dignity for all citizens' (Helstosky, 2006, p133). As a result of this political approach, the Italian Government took up the study of food consumption habits as important markers of the nation's socioeconomic progress. The National Institute of Nutrition, in particular, carefully studied the initiatives under way and found out, for example, that institutional meals and snacks were providing workers and schoolchildren with 40–100 grams of protein per day and that the city of Milan was feeding schoolchildren supplemental snacks of between 700 and 1400 calories.

The analysis of data collected at the time showed that school feeding 'contributed dramatically to good health and steady development among children, so much so that many studies concluded that post-war food programmes might possibly reverse the effects of wartime malnutrition among children' (Helstosky, 2006, p132). It was at that point that many of the early school cafeterias were set up in Italy, usually through a joint effort by the Administration for the Italian and International Welfare Activities (which was part of the United Nations Relief and Rehabilitation Administration), the Food and Agriculture Organization (FAO) of the United Nations and the National Institute of Nutrition.

In the following decades, the school meal service was increasingly utilized by Italian public authorities as a tool to reinforce healthy food habits and, more generally, to create new generations of knowledgeable and informed consumers. Through an analysis of a national food education programme, in this chapter we will distil the values Italians attach to school food. We will then examine how these values have informed the school food revolution recently implemented in the city of Rome, where creative public procurement policies are designing a food system that has the highest potential to deliver the wider environmental, economic and social benefits of sustainable development.

School Meals as a Health and Educational Service: The Italian Model

Whereas countries like the UK and the US have for decades treated the school meal service as a commercial service, in Italy school meals are recognized as an integral part of people's right to education and of consumers' right to health. The legal analysis provided by Italian law professor Ugo Ruffolo (2001, pp104–105) is an ideal starting point to distil the vision of the school meals system that distinguishes Italy from most other countries in the world.

For Ruffolo, there are as many as five articles of the country's Constitution that, albeit indirectly, guarantee children's right to local and healthy food. These are:

1. the *fundamental right to health* to which each individual is entitled (Article 32);
2. the *inviolable right to a harmonious personal development* that each citizen has, *both as an individual and as a member of social groups* (Article 2);
3. the promotion of *cultural and territorial development* (Article 9);
4. the *protection of children*, both as family members (Articles 29, 30 and 31) and as members of territorially-based social collectivities, such as schools (Articles 33, 34 and 37); and
5. the *valorization of local autonomies* and the emphasis on devolution of decision-making powers (Articles 5 and 114).

Within this legal context, school meals in Italy are assigned a dual educational function: on the one hand, they are expected to teach schoolchildren the values of territoriality and local traditions; on the other, they must help them to acquire a sense of taste that would contribute to their personal development.

This vision shapes the Italian school meals system in three fundamental, and very peculiar, ways (see Morgan and Sonnino, 2005 and 2007). First, by embedding school meals in a much broader educational project, the Italian system allows contractors to retain complete control over the service. In fact, the law establishes that the contractor is entitled to an *ius variandi*, or the right to modify the service agreed in the contract if changes are introduced in the wider educational programme of which school meals are part (Article 1661 of the Civil Code). Furthermore, the law establishes that the contractor has the right to monitor the school meal service to ascertain whether it conforms to the educational and cultural parameters specified in the contract (Article 1662 of the Civil Code).

Second, by assigning to school meals the function of conserving local traditions, the Italian system legitimates the possibility of 'discriminating' – that is, of privileging local operators and all expertise linked to local food. As stated by the State Council in 1992, it is legal for a municipality to restrict the participation in a public competition to companies located in the province, 'given the necessity to take into consideration the taste of local consumers and to guarantee prompt communication and intervention in case of problems' (Cons. Stato, V, 24 /11/1992, no 1375, in *Cons. Stato*, 1992, p1636, cited in Ruffolo, 2001). This, in turn, opens up a legal way to overcome the European principle of 'non-discrimination' we discussed in Chapter 2.

Third, in Italy the awarding of catering contracts is based on a notion of 'best value' that is interpreted and evaluated by taking into consideration not just economic issues (the proposed price of the meal), but also the hygienic, nutritional and organoleptic aspects of the service, as well as its compatibility with the wider educational context. In Ruffolo's words (2001, p117), 'the economic advantage [...] is appreciated not only on the basis of the criterion of the "lowest price"; the qualitative characteristics of the proposed service in terms of "food culture" and its compatibility with the curriculum are also taken into account.'

This interpretation of the school meal service has enabled Italian public authorities to design and implement a variety of healthy eating policies, which have emerged at different levels of governance. In fact, diet in Italy is considered as a 'concurrent' subject: the general guidelines are provided by the National Government, but their implementation is a responsibility of the Regions. In a country characterized by a high level of local political autonomy, some municipalities have also independently designed their own school meals systems. As a representative of the General Directorate for the Quality of Agri-Food Products stated:

> *Regions and municipalities have taken autonomous initiatives in this field, so at the moment we have a jungle of regional and municipal regulations; there are even sentences of the Regional Administrative Courts attempting to impose organic products in schools – it's a jungle!*

Historically, the earliest initiatives in the field of sustainable public food procurement go back to the mid-1980s, when the National Institute for Nutrition published its *Guidelines for a Healthy Italian Diet* (1986), which promoted the Mediterranean food model in public sector catering (Soil Association, 2003, p63). At the same time, the town of Cesena, in Emilia-

Romagna, implemented the first organic school meals system. Between 1989 and 1990, two other significant experiments took place: the city of Padova, in Veneto, created the first organic university canteen, and Udine, in Friuli-Venezia Giulia, introduced an organic hospital menu.

During the 1990s, a general concern over the health impacts of BSE and pesticide residues further reinforced the Italian commitment to food education and local sourcing (Morgan and Sonnino, 2005). While the media were portraying BSE as something foreign, 'coming from a different country where people do not know how to eat and how to farm' (Sassatelli and Scott, 2001, p225), the Mediterranean diet became an ideal tool 'to move meals away from mass produced, highly processed and invisibly adulterated foods of unknown provenance towards the use of more whole foods and a greater proportion of certified organic ingredients' (Soil Association, 2003, pp63–64).

In response to this new awareness, in December 1999 the Italian Government issued Finance Law 488, a very innovative piece of legislation that establishes a direct and explicit link between organic and local food and public sector catering. As stated in Section 4 of Chapter 1, 'Measures to facilitate the development of employment and the economy':

> *To guarantee the promotion of organic agricultural production of 'quality' food products, public institutions that operate school and hospital canteens will provide in the daily diet the use of organic, typical and traditional products as well as those from denominated areas, taking into account the guidelines and other recommendations of the National Institute of Nutrition.* (cited in Soil Association, 2003, p65)

This national law created a regulatory context that encouraged many municipalities to turn organic, especially, but not exclusively, in the six regions[1] that at that time had issued specific laws to promote the use of organic products in public canteens. Within a decade, the number of organic school canteens grew from 70 (in 1996) to 658 (in 2006), while the total number of organic school meals served annually nationwide increased from 24,000 (in 1996) to 1 million (in 2006) (Bertino, 2006).

Along with organic products, which Italian caterers tend to prioritize for health reasons (Bertino, 2006), more recently Italian procurement officers have also started emphasizing local foods. Law 488/1999 explicitly promotes the territoriality of the school meals system by establishing that the quality of the proposed services must be assessed by also considering the relationships that school food has with local cultures and traditions. As a result, municipalities such as Fanano (Emilia-Romagna), Ascoli (Marches) and Borgo San Lorenzo (Tuscany) have decided to utilize exclusively local produce for their school meals (VITA Non-Profit Online, 2003). In some cases, initiatives are under way to re-localize the organic school food chain. In Budoia, a small town in Friuli-Venezia Giulia, after discovering that the organic cabbage served in the schools was sourced from The Netherlands, parents successfully took the initiative to mobilize local organic producers in their area (Green Planet.Net, 2006). The initiative set in motion a peculiar model of public food procurement that has been at the centre of a legal controversy in Italy (see Box 4.1)

Box 4.1 *A family's business: The Budoia school meals model*

Budoia is a small town of 3000 people located in the region of Friuli-Venezia Giulia. Although it provides meals for just 140 children, Budoia has become well known in Italy for the peculiarity of its school meals system.

As a response to children's dissatisfaction with the meals provided by a multinational corporation, which used to prepare the food in cooking centres located 15km from the school, in 2005 parents decided to form a cooperative and take over the service. Today, one of the mothers is in charge of purchasing the food, one father supervises the kitchen and another mother tastes the food and makes sure that children like it. Through the support of the Italian Association for Organic Farming, the cooperative has managed to provide school meals based on 90 per cent organic ingredients and more than 50 per cent local ingredients.

Once a week, the parent in charge of choosing the ingredients purchases whatever is available fresh on the market and designs the menu accordingly. The Council, which sees this system as a strategy 'to support local agriculture', as the Mayor stated, pays for the food without going through a tendering process. For this reason, in 2006 the Council was sued for 'administrative irregularities'. Significantly, the Court decided that 'public interest' took precedence and the case was brought to an end.

As this brief history of the school meal service demonstrates, Italian food procurement policies are primarily informed by values of safety, seasonality and territoriality that set up an 'alternative' metric for the evaluation of the bids submitted by suppliers and catering companies. Significantly, these are also the values that public authorities attempt to transfer to future generations through a wide variety of food education initiatives.

In Italy, the 'whole-school' approach is an integral part of the school food service. For this reason, there is a wide range of education initiatives implemented in schools. In 1992, for example, Slow Food launched a programme, called 'Week of Taste', which turned Italian classrooms into workshops and kitchens to involve the youngsters in supervised tasting and practical food experiences (Petrini, 2001). In 1998, Slow Food published *Dire Fare Gustare: Percorsi di Educazione del Gusto nella Scuola* (*Talking, Doing, Tasting: Taste*

Education in Schools), a book for teachers and parents that suggests experiments to teach children how to use their senses, how to learn about raw ingredients and how to rediscover conviviality. According to Petrini (2001, p76), the founder of Slow Food:

> *Learning the sensory alphabet, then constructing the grammar and syntax of taste allows the student to come to an understanding of the cultural values tied to […] local specificity. Thus food is linked to history, to social processes, […] transcending the simple idea that eating is just swallowing. In the end, [...] aware consumers will come into being.*

More recently, Italian food culture has become the focus of a national educational programme called 'Cultura che Nutre' ('Culture that Feeds'). In the next section, we will analyse the key aspects of this initiative, which, we argue, provides an ideal context to identify the cultural and educational values embedded in the Italian school meals system.

Culture that Feeds, Feeding a Culture: The Cultural Dimension of Sustainable Food Procurement

Cultura che Nutre is part of an inter-institutional programme called 'Food Education and Communication', which was launched in 1998 by the Ministry of Agricultural and Forestry Policies and by the Regions to create a network among different actors involved in food education. At the central level, the programme is coordinated by an inter-institutional committee formed by representatives of the Ministry of Agricultural and Forestry Policies, the Regions, the Ministry of Public Education, INRAN (the National Institute for Research on Diet and Nutrition) and ISMEA (the Institute for Study, Research and Information on Agricultural Markets). Broadly speaking, the Food Education and Communication programme has nine objectives (Finocchiaro, 2001, p16):

1 to develop Italian agriculture through the promotion of quality regional products and the valorization of rural traditions;
2 to disseminate the principles of a healthy diet in which food is seen as 'symbol, culture, history, respect for the environment and knowledge of the Italian agri-food patrimony and territory';
3 to promote food education as education for wellbeing and informed consumption;
4 to involve consumers in the monitoring of food quality;

5 to create synergies in the field of food education;
6 to promote scientifically sound dissemination initiatives about food;
7 to create responsible consumers;
8 to promote taste education; and
9 to disseminate information about consumers' rights.

At the regional level, Cultura che Nutre is administered by committees that comprise local, provincial and regional representatives. As mentioned above, Italian regions have a certain degree of autonomy in implementing diet-related initiatives. A representative of ISMEA explained:

> *Every region can personalize the programme, as long as it complies with the general guidelines. So, for example, some regions organize training programmes for the teachers, others target mostly dieticians and nutritionists, still others involve both. There is a shared framework and then there are regional interpretations.*

Cultura che Nutre is one of the most imaginative strategies implemented so far to reach the objectives of the Food Education and Communication initiative. Introduced in all Italian regions, the programme aims to educate schoolchildren – who, as one of the coordinators of the programme explained, are seen both as 'future consumers' and as 'influential messengers' to their families – to adopt an 'informed and healthy' diet.

The educational material distributed to school teachers to help them organize their class activities consists of three books that focus on different dimensions of food (see Morgan and Sonnino, 2005). The first book, *The Land, the Product, the Market*, analyses the economic aspects of food. After a brief discussion of the multifunctional role of agriculture in Italy, the book explores at length the concept of food quality, which is presented as embracing simultaneously agricultural, health, nutritional, organoleptic and socio-psychological attributes. This holistic notion of quality is a key aspect of the programme. As the ISMEA representative stated:

> *In Italy food education has always been approached from the health perspective, from a scientific perspective, from a nutritional perspective. The methodology [of Cultura che Nutre] is new [...]. In addition to health, we have tried to approach food education from a sociological, anthropological, psychological and folkloric standpoint. This integrates perfectly with other policies that focus on the conservation of the Mediterranean tradition, of Italian typical products and of traditional regional cuisines.*

In its last section, the book provides teachers with suggestions on how to practically implement food education at different school levels. For nursery school children, the book recommends a trip to the countryside where children can learn to identify plants, animals, foods and flowers. For elementary school children, teachers are advised to explore in class the most common methods of food conservation through practical experiments on dairy products. The proposed research activities for secondary and high schools focus, respectively, on herbs and medical plants and on Italian typical products.

The second book, *I Eat Therefore I Am*, analyses nutritional habits in relation to wellbeing, health and lifestyle. Teachers are told about the changes that have occurred in Italian food habits, the links between diet and health, and the social costs of an unhealthy diet. The class projects that have been designed to educate children on these issues vary according to their age. Elementary school teachers are encouraged to design research activities on the history, geography, science, preparation and legislation surrounding milk and dairy products. The objective of the project proposed for secondary school children is 'informed consumption': pupils are stimulated to classify food on the basis of its nutritional function, its chemical composition, its methods of preparation, its seasonality and its traceability. For high school students the book proposes an empirical investigation of food costs, including production, distribution, advertising, stocking, processing and preparation.

The third book, *At the Table with Taste and Culture*, explores the historical, anthropological, psychological and sociological aspects of Italian diet. Teachers are provided with information on the relationships among food, socialization, traditions, identity, personal development and history. The proposed research projects to be carried out in the schools focus on the analysis of taste (for nursery and primary school children), on the relation between food and the five senses (for secondary school children) and on the relation between food and social trends (for high school children).

Schoolchildren participating in Cultura che Nutre also receive some printed materials. Again, the emphasis is generally on the two most fundamental values Italians attach to food: seasonality (Figure 4.1) and territoriality (Figure 4.2). Primary school children receive two books that focus on these values: *The Land of Good Things*, which emphasizes the meaning of seasonal food, and *The Festival of Good Things*, which describes the most renowned certified food products in Italy. In addition, they also receive a videotape, divided into 12 audio-visual units, which, as the methodological coordinator of the programme stated, aim to 'strengthen children's links with their roots and territory, promote their understanding of agriculture as a fundamental tool for conserving territorial resources and enhance their awareness of the relations existing among productive systems, food systems and environmental protection' (Finocchiaro, 2001, p18).

Figure 4.1 *Seasonality*

To teach children the value of seasonality, the book *The Land of Good Things* includes various games. Here children are asked to colour the fruit and vegetables illustrated and to draw lines connecting each of them to the appropriate seasonal basket.

Increasing children's knowledge of typical and healthy foods is the goal of other initiatives proposed by Cultura che Nutre. In September 2003, a 'sensory' laboratory was installed at the International Fair of Natural Food, Health and the Environment to teach children about typical agri-food products. The same year, a national competition among schools was launched to reward with a holiday on an agri-tourist farm the 20 classes (one for each region) that created the best 'Snakes and Ladders' game devoted to a typical product from their own region. In 2006, Cultura che Nutre proposed theatre shows representing a battle between vegetables and mass food products – a battle, of course, won by the healthy and fresh vegetables.

In short, the Italian system of public food procurement is a product, rather than a cause, of a deeply embedded culture that connects school meals (and food in general) to local identity. Through creative procurement strategies and the implementation of educational programmes such as Cultura che Nutre, the Italian Government is creating knowledgeable consumers willing and able to

In Sardegna non c'è solo il mare. Ci sono anche un sacco di pecore. Anzi, da sole sono più della metà degli animali dell'isola. Questo spiega bene il formaggio pecorino, uno dei motivi di vanto dei sardi. Le pecore fanno il latte, i pastori con il latte ci fanno il pecorino. Sfortunatamente per le pecore, l'uomo è anche carnivoro, e la carne di pecora e di agnello è molto ambita, specie in certi periodi dell'anno. Così, un altro vanto di questa terra bellissima è l'**Agnello di Sardegna**, o meglio la sua carne.

Gli agnelli di Sardegna devono essere allevati allo stato brado.

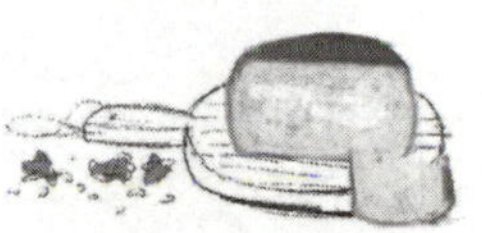

Il formaggio pecorino è un altro prodotto protetto.

Piace tanto che a volte certi commercianti disonesti hanno venduto carne di agnelli polacchi o rumeni fingendo che fossero sardi. Ma il marchio IGP, se ricordate, serve proprio a evitare confusioni e disonestà. Qui per esempio il marchio garantisce che quell'agnello è un tipo a posto, che viene dalla Sardegna, ha preso il sole e il vento dell'isola, ha scorrazzato liberamente sui suoi prati e bevuto la sua acqua, ha succhiato il latte della mamma e ha mangiato solo le erbette del posto. Più sardo di così!

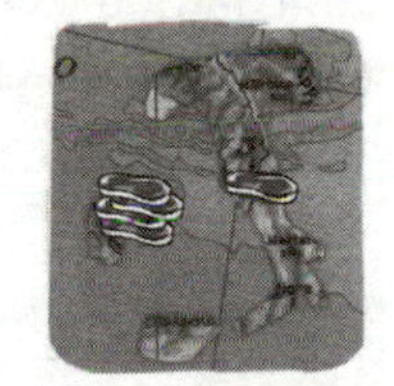

In Sardegna si mangia carne quattro volte di più che nel resto d'Italia.

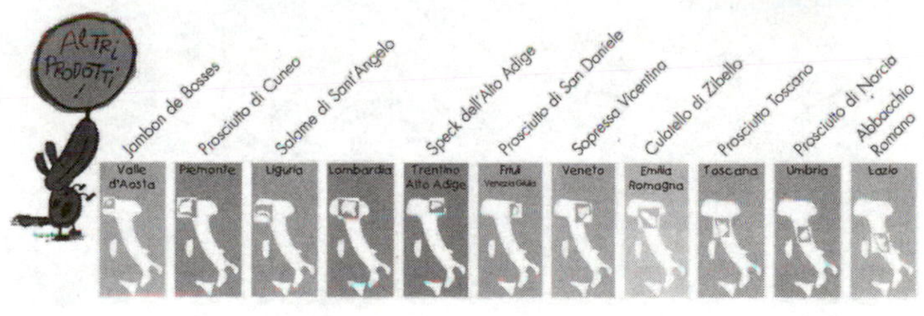

Figure 4.2 *Territoriality*

The book *The Festival of Good Things* teaches children about typical and certified products and their links with the area of production. The main text here focuses on Sardinia's sheep and the products obtained from them (pecorino cheese and meat). Special attention is devoted in the text to the meaning of the Protected Geographical Indication (PGI) label that distinguishes Sardinian lamb. It is explained that the label guarantees that the animal was raised in the island and in the wild. At the bottom of the page, children are reminded of the links existing between typical products and specific Italian regions.

sustain the 'local'. But this is not a form of defensive localism. As the ISMEA representative explained:

> *You would not be blaming a school for teaching the literature produced in its own country. In Italy I teach Italian literature. [...]*

> *And it is the same with consumption. [...] If one learns to appreciate Dante, it is more likely that one will also read Cervantes.*

Significantly, the main story in this chapter comes from a city, Rome, which has only recently started to re-localize its school food chain. As we will describe in the next section, in Rome 'quality' means many different things: health, ecological sustainability and, perhaps most important, social justice. It is on the basis of these values that Rome has implemented a school food revolution on a scale never seen before.

School Meals in Rome: The Quality Revolution

When Law 488/99 was issued, Rome was governed by a Green Party administration, led by Mayor Francesco Rutelli, which was especially interested in the potential of organic catering in schools. However, for a city that feeds nearly 150,000 schoolchildren every day for 190 days a year, serving almost 150 tonnes of food per day (see Figure 4.3), the prospects of an organic conversion raised peculiar challenges.

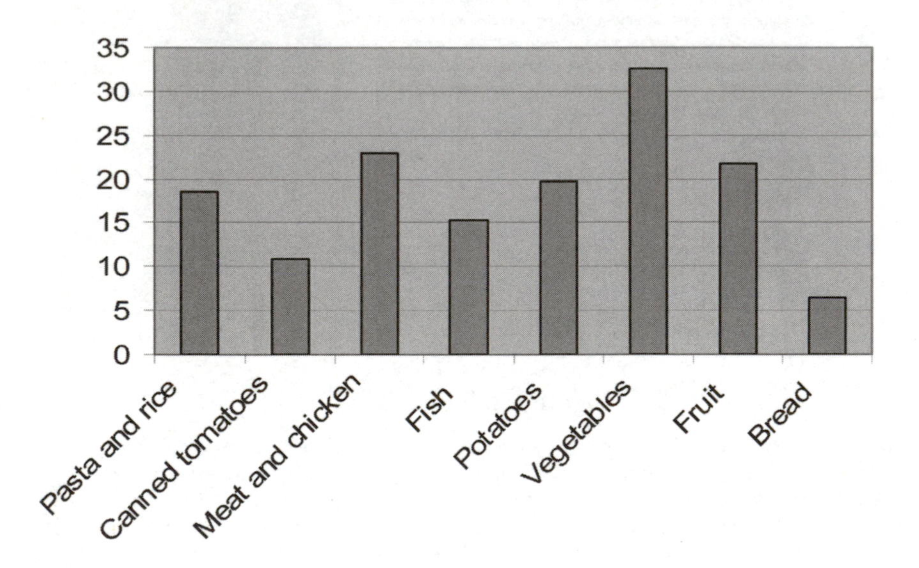

Figure 4.3 *Amount of food ingredients served daily in Rome's schools (tonnes)*

Source: City of Rome (Smargiassi, 2007)

As Silvana Sari, the Director of the Department for Education in charge of school meals, explained:

> *We were the largest contractor in Italy, 40 per cent of all public school meals were catered in Rome – and we knew that our demand could have raised the prices of organic products in an uncontrollable way.*

Determined to revolutionize its school meals system, Rome chose a progressive procurement approach. On the production side, representatives from the organic certification bodies were asked to identify the products ready to sustain the economic impacts of Rome's massive demand. On the consumption side, nutritionists were consulted to find out which organic products are most beneficial to children's health. From both ends of the food chain, Rome received the same response: fruit and vegetables needed to be prioritized (Sonnino, 2009).

Rome authorities decided that the implementation of a quality revolution also required an upheaval in the city's procurement policy and tendering procedures. In the Italian procurement systems, local authorities play a central role, as they are delegated the fundamental responsibility of designing the tender through which school catering companies are selected. In a large city such as Rome, however, an important role is also played by the districts, or *Municipi*, into which the capital is divided. In fact, the *Municipi* collaborate with the central Department for Education in drawing up the tendering specifications, they provide advice on the nutritional aspects of school food on the basis of their individual experiences, they are responsible for some of the inspections and they directly manage the school meal service offered by nursery schools. Rome's governance system, then, is not as centralized as New York City's, nor is it as decentralized as the London governance system we will describe in the next chapter. Its most significant tensions tend to develop particularly between the *Municipi* that are governed by political parties of the opposition and Rome's central departments.

During the ten years that preceded the beginning of its quality revolution, Rome had adopted a 'negotiated procedure' through which a selected number of catering companies were invited to tender for the eight territorial lots into which the 20 *Municipi* had been subdivided. Under this system, Sari pointed out, the school meal service became characterized by 'very high prices, no rules, basically no organic products – and no control over the number of staff needed to guarantee a quality service'. To change this situation, Rome decided to adopt an 'open procedure', under which all catering companies interested in supplying Roman schools were given an opportunity to respond to the call for tender. At the same time, to break the consolidated linkages that had developed amongst Rome's catering companies, the city's territory was redivided into 11 lots.

Significantly, Rome's authorities did not consider 'quality' and 'price' as irreconcilable goals. As Sari stated:

> *If a public administration knows what it wants and clearly identifies the various components that make quality, it can easily predetermine everything and it can foster a competition based on the maximum rebate.*[2]

As in other European countries, the contracts were indeed awarded on the basis of the 'economically most advantageous tender'. However, in contrast, for example, with the UK, this principle was not interpreted in terms of cost reduction alone. Rome also took into consideration the socio-environmental externalities of the proposed service in its evaluation of the bids submitted (Sonnino, 2009). Specifically, the 2002–2004 tender identified a number of essential criteria that aimed to guarantee the basic quality of the service. Under the paramount priority of protecting children's health, catering companies were required to provide fresh organic fruit and vegetables during the first year of the contract and to add organic legumes, bread, baked products, pasta, rice, eggs and canned tomatoes during the second year. An exception was made for vegetables with a short harvesting season, such as peas, green beans and spinach, which could be supplied frozen (Comune di Roma, 2001).

In addition, the tender introduced a set of very innovative award criteria that aimed to stimulate bidders to further develop the socio-environmental quality of the products and services offered. Contracts were awarded on a 100-point award system in which the price proposed accounted for 51 points. Another 30 points were awarded on the basis of the organization of the service (the number of staff and working hours offered by the catering companies, the environmental certifications they held and the environmental friendliness of their transportation system). 15 points were awarded for 'projects, interventions and services' proposed to promote food education amongst the users of the service[3] and to reduce noise in selected school canteens. Finally, 4 points were allocated to catering companies capable of offering additional organic, PDO and PGI food products – beyond those required in the tender (Comune di Roma, 2001).[4] By 'leaving to the market itself the opportunity to respond', Sari stated, Rome unexpectedly managed to also bring to the children's table organic olive oil, mozzarella, yoghurt, veal, pork, turkey and ham (Sonnino, 2007a).

A 'Roman model, based on food security and quality and on the idea that the meal consumed at school is an educational experience', as Sari described it, began to emerge, but not without difficulties. On the production side, catering companies had been struggling for years to respond to the national demand for organic food in the public canteens. As the director of one catering company explained:

> *When we started dealing with organic products in the late 1990s, before the Finance Law, there were periods of the year in which*

> *some products were not available or you had to go back to the source to get them, which implied also paying for the higher transportation costs. These difficulties and costly activities were not rewarded by local authorities.*

The new demand from Rome further complicated this scenario. Another caterer said:

> *Today we are asked [by Rome] to provide not a dish of the day, but a level of quality that is good and that remains consistent. So we cannot plan our activities on a daily basis – you can't do that when you're dealing with these numbers.*

Faced with all the economic and logistical difficulties that supplying such a large market involved, the contracted companies requested and obtained a dialogue with Rome's authorities. A permanent round table was created to allow public institutions, producers and suppliers to meet on a regular basis to discuss problems, to do the necessary planning and, perhaps most important, to foster 'a shared willingness of going in a certain direction', as the director of one catering company explained (Sonnino, 2009).

At the same time, however, Rome began to develop a very stringent control and monitoring system that left no room for manoeuvre to suppliers involved in the city's quality revolution. All contracted companies were required to obtain the ISO 9001 quality certification[5], to develop an HACCP plan[6] and to produce a handbook of good hygienic practice specifying all the rules adopted with regard to the hygiene of staff, equipment and premises. In terms of control, it was established that contractors' compliance with the guidelines specified by the city could be verified by a number of different bodies. The *Municipi* manage the contracts and the relationships with the contracted companies. Through their dieticians, they monitor the service and apply sanctions and fines in case of misconduct. The Central Department for Education supports and advises the *Municipi* but can also perform autonomous inspections through its dieticians. There is also a specialized contracted company hired by the City to verify the quality and hygiene of the food, the hygienic and cleaning procedures utilized in the premises, the organization of labour, compliance with the menus and weights established by the City, the certification of organic products, the maintenance of equipment and machinery, and the implementation of the HACCP (Sonnino, 2009).

In addition, local health authorities can inspect the schools, which are also individually monitored by so-called 'Canteen Commissions'. Formed by at least two parents and present in every school, the Canteen Commissions assess compliance with the menus, the expiration dates of the food products and the hygienic conditions of the premises (Comune di Roma, 2004b, pp23–27). The participation of the Canteen Commissions in the school meals system was crucial to address problems that emerged at the consumption end of the food chain. In fact, it enabled the city to maintain a constant dialogue with the users of the service, which was especially important at the onset of Rome's quality revolution. As Maria Coscia, the former Councillor for Education, recalled:

> *Back then parents were sceptical, because in Italy people were used to food products that looked good. Organic products are small and at times they look rotten, so there was an initial lack of trust amongst consumers.*

In 2004, Rome entered the second stage of its school food reform. In the name, again, of children's health and food security, the 2004–2007 tender reduced the amount of animal protein to be contained in meals; specified the exact weight of all cooked food; defined the modalities that catering companies must adopt with regard to the conservation, handling, cooking and distribution of food; introduced healthy mid-morning snacks (fruit tarts, bananas and bread rolls) for all schoolchildren; diversified meals around children's age; and redesigned the menus on the basis of four quality principles:

1 *Seasonality*: The menus, which have a summer and a winter version, are mostly based on the use of fresh foods.
2 *Variety*: To ensure an adequate intake of all necessary nutrients and introduce children to different foods, it was established that no dish is to be served more than once every five weeks.
3 *Territoriality*: To emphasize the links between food and a specific area of production, Rome prioritized certified meat products – a move that opened up a new market not only for Italian veal producers but also for Welsh lamb producers and French pork producers. In addition, it was established that the bread served in schools must be baked and packaged within six hours and must be consumed no longer than 12 hours after packaging.
4 *Nutritiousness*: On the basis of the guidelines provided by the Italian Institute of Nutrition, Rome established that the mid-morning snack must provide 8–10 per cent of the recommended daily nutritional intake for children, while lunch must guarantee 35 per cent. In addition, special menus were designed to guarantee a healthy diet for the roughly 4500 children who have dietary restrictions for medical or religious reasons (Comune di Roma, 2004b).

In addition to emphasizing children's health, the tender also introduced new incentives to improve the social and environmental sustainability of Rome's school meal service. The list of mandatory organic products expanded considerably to include all the products that catering companies had offered as optional with the first tender. Again, caterers capable of offering additional organic products were awarded four points – an initiative that added organic Parmesan, mozzarella and butter to Rome's school food and has enabled the city to achieve a total of 70 per cent organic supply to its schools[7] (Comune di Roma, 2004b).

To reduce the potential for contamination of organic products and, at the same time, to develop short supply chains, Rome rewarded with another four points catering companies able to supply 'bio-dedicated' products – in other words, foods that are produced, processed, packaged and distributed by firms that operate exclusively in the organic sector. To emphasize provenance and traceability, nine points were allocated to bidders that offered PDO and PGI products. Finally, under the stated objective of 'preventing situations in which the actions we perform to improve our quality of life threaten the quality of life of people in other areas of the world or that of future generations', the City allocated two award points to encourage catering companies to source Fair Trade products (Comune di Roma, 2004b). As the Councillor for Education explained:

> *In the schools we organize many activities around the theme of solidarity with developing countries. [...] The meal is [...] an educational moment; it's not just a question of feeding the children, it's part of a broader project.*

Today, 280,000 Fair Trade bananas from Ecuador and 140,000 Fair Trade chocolate bars from the Dominican Republic are served every week in the Roman public schools – a figure that has increased by 20 per cent the market for Fair Trade products in Italy[8] (Massimiani, 2006, p17). With regard to the service itself, Rome decided to reward initiatives proposed to improve the kitchens and the eating environment (17 points) and to employ a high number of staff as well as trained dieticians (5 points). Finally, 8 points were awarded to projects proposed to educate parents and teachers on food issues (Comune di Roma, 2004b).

The food system that Rome is creating is in many ways blurring the boundaries between the conventional and alternative food sectors. On the Roman schoolchildren's plate, organic, bio-dedicated and certified foods coexist with more conventional products (Sonnino, 2009). As one caterer explained:

> *The critical point in the food chain concerns self-control, the application of the HACCP. Large companies always have labs that perform microbiological and chemical analyses on the products they*

purchase and then sell to us. [...] It is difficult to retain the artisanal dimension in all of this.

Furthermore, artisanal productions cannot reach the capacity needed to supply a large urban system:

The artisanal producer cannot provide a standardized product in this kind of volume. With artisanal producers you can reach high levels of excellence, but there is a total lack of standardization. [...] Conventional products in Italy have an average qualitative standard that is not simply acceptable, it is good, and it meets the standards established by the tender. [...] With bio-dedicated, Fair Trade, and PDO and PGI products there are problems, in the sense that it is a brand new market that only exists in Rome and in a few other places. The supply does not meet the demand.

This food system is not without weaknesses, especially with regard to its environmental sustainability. For example, some products leave the Roman countryside, where they are produced, to reach national food platforms that are located elsewhere in Italy and then come back to Rome. For the catering companies, these food miles are unavoidable:

It is certainly a waste of energy, but this is the only way to achieve the kind of volumes we need to supply Rome.

In addition, producers and caterers had to face considerable financial difficulties. In commenting on the price of €4.23 per meal[9] that the city was paying until 2007, the director of one of the contracted companies said:

We have been lucky to be able to accompany them along this path of qualification, of specialization, but catering companies have their own adjusting time, which is different from that of the public administration. [...] If they don't raise the price, today's excellence can become tomorrow's weakness.

For another caterer:

They must understand that this guaranteed, verified and verifiable quality has a price. [...] Without its size and its visibility on the market, Rome would have never achieved this level of quality at the price it pays.

Once again, Rome decided to listen to its suppliers. As Sari explained:

> *Through the whole process, we have created a new respect for the public customer. In the past, catering companies were in control. [...] With the second tender [...] competition has been accepted. The catering companies have realized that if they promise something [...] they are expected to deliver. They have understood that the City of Rome is capable of managing the tender and wants to manage it. And they have given up on ideological and political attacks. [...] At this point it is important that we listen, because we have asked them to make an enormous financial effort.*

In March 2007 the city entered the third stage of its quality revolution (Comune di Roma, 2007). The new tender covers five, rather than three, years – a time span designed to allow producers and caterers to adjust to the new demands and to make new investments. The price paid to the contracted companies has been increased to €5.28 per meal. In exchange, contractors have been asked to develop further innovations. Once a month, they serve ethnic dishes to celebrate the increasing diversity of the Roman school population. Leftovers from school lunches are delivered to animal shelters, while unutilized foods are destined for charity associations. Incentives are provided for catering companies that source food from social cooperatives that employ former criminals or that work land confiscated from the mafia. In response to growing environmental problems, the providers of the school meal service are required to utilize only biodegradable plates, to develop recycling schemes and to utilize detergents that have a low environmental impact (Maisto, 2007).

Significantly, amongst the new award criteria identified by the City of Rome there is also *guaranteed freshness*, which rewards the utilization of food products harvested no longer than three days before being served in the schools. To assess the freshness of the foods offered, Rome is the first city to look at food miles – or, specifically, the number of kilometres and hours that separate harvesting and consumption.

After years of efforts and continuous improvements, Rome's reform process has perhaps reached a stage of non-return. As Sari stated:

> *We have created so much cultural awareness that everybody now accepts and understands that quality in Rome will never be sacrificed again.*

This time, caterers seem to be ready for the new quality requirements. Quoting again the director of one of the largest catering companies:

> *Rome has a strong emphasis on quality [...] in the widest sense of the term, starting from the quality of foodstuffs in terms of taste to the quality of foodstuff in terms of provenance. [...] There has been a progressive evolution. [...] And there is also a broader type of quality that concerns the service, so not just what is eaten but also how the food is prepared. And then there is an attention to the environment in which the meal takes place.*

Public Procurement and Social Inclusion: The Roman Model

As described in the previous section, Rome's public procurement approach has three main characteristics (Sonnino, 2009). First, it is a *dynamic* approach that constantly renegotiates, and progressively revises, the notion of food quality. At the onset of the reform, the city's procurement team explained, Rome identified the improvement of children's health and safety as the paramount goal of its school food revolution (Comune di Roma, 2004a). In this context, organic and 'bio-dedicated' products have been prioritized because of the absence of pesticide residues.

However, as previously noted, the organic market at the time had not yet reached the capacity to provide 27 million school meals a year. To deal with this issue, Rome decided, on the one hand, to adopt an incremental approach to the organic conversion of school food and, on the other, to broaden its interpretation of quality through an emphasis not just on the 'naturalness' of the products offered, but also on their provenance.

Significantly, food provenance in Rome is emphasized at both the local and the global level. When asked to explain the requirement for PGI meat, for example, the head of procurement referred to health reasons (specifically, the higher nutritional value provided by certified animal breeds), but he also pointed out that, given the paucity of certified meats in Europe, the decision was made to create new opportunities for local meat producers. In this sense, Rome's emphasis on provenance becomes primarily a re-localization strategy, a tool used to promote local economic development and environmental protection, as also implied by the new requirement of 'guaranteed freshness'. At the same time, however, in Rome provenance is also a strategy to open up new opportunities for local producers in other areas of the world. Through its demand for Fair Trade products, the city in fact promotes social justice and solidarity beyond its most immediate boundaries.

Second, and partly as a result of all of this, Rome has adopted an *integrated* procurement approach that is able to reconcile different (when not conflicting) quality conventions. The four 'worlds of justification' discussed in the agri-food literature (see Chapter 1) coexist in the Roman school meals system.

Indeed, *market conventions* guided at least two fundamental decisions made at the beginning of the reform: the identification of the price proposed as the single most important criterion for the awarding of the contracts and the radical changes introduced in the tendering procedures, with the shift to a system that opened up the competition to all suppliers interested in participating – a move that has proven crucial to stimulating the organic market.

Issues of standardization and reliability, which lie at the core of *industrial conventions*, are emphasized through expectations in terms of the environmental certifications that all contracted companies must possess and through standardized quality controls across the food chain. At the same time, the massive demand coming from such a large city inevitably implies a more general standardization of the quality of the foodstuffs themselves, which in practice means an emphasis on conventional products at the expenses of artisanal techniques of production, as caterers pointed out. In this regard, the analysis of Rome's school food reform may prove, as Hatanaka et al (2006) argue, that the production and maintenance of quality in the food sector require both standardization and differentiation. As we have seen, the latter is in fact also promoted in Rome through a progressive demand for local (bio-dedicated and certified) products.

Combined with the development of menus and food education initiatives revolving around the values of territoriality, the demand for local products embeds in the Roman school meals system the *domestic conventions* of attachment to place and traditional production methods. Through the introduction of Fair Trade products and ethnic dishes, these domestic conventions become inextricably linked with *civic conventions* that extol the socio-cultural meanings of food and its capacity to foster social justice and solidarity across space – in other words, towards nearby foreign fellow consumers and distant foreign producers.

Third, Rome has adopted an *inclusive* approach that has actively involved in the reform process both producers and consumers. On the supply side, the establishment of a permanent round table where catering companies regularly meet with city authorities and procurement officers has created relationships of trust and mutual respect between the public and private sectors. It is mostly because of this trust and respect that caterers have managed to endure, during the early years of the reform, the difficulties of entering a market that initially left them with a profit margin of just €0.13 per meal served in the schools. Quoting the example of a three-month 'moratorium' that the city granted to the providers of the service to allow them the time to organize themselves, the director of one of the contracted companies stated that the City authorities 'have been tolerant, and the catering companies have been responding to this positively. We feel they understand us.'

At the same time, consumers have also been included in the reform of the service. In general, children and teachers are involved in food education initiatives organized by caterers, which, as noted above, aim to share with the users of the service all main aspects of Rome's school food revolution – from its technicalities to the values that underlie it. Parents, on their part, are given access to the system through the operation of the Canteen Commissions, which, as explained above, play a role in the monitoring and control of school meals.

More specifically, Rome's inclusive approach manifests itself through the initiatives taken to incorporate in the system children with different socioeconomic and ethnic backgrounds. In fact, the city has made the school meal service free for families that have an annual income of less than €5164, while families with an income of less than €12,911 are entitled to a 25 per cent discount on the monthly fee. In the name of social justice, since the very beginning of the reform, Rome has also designed special menus for the roughly 4500 schoolchildren who have specific requirements related to religion or health. More recently, as mentioned earlier, the process of social inclusion has been further broadened through the introduction of ethnic dishes designed to celebrate the increasing cultural diversity of the city's school population and to use food as a means to promote the value of solidarity (Maisto, 2007).

Quality Food and Social Justice: Some Conclusions

The Roman school meals system has not emerged in a political vacuum. Quite the contrary, it has been shaped and supported by a governance philosophy that emphasizes the inextricability of economic development and social cohesion. For many observers, this philosophy has recently shaped a distinctive and very effective 'Roman model', which holds the potential to create a new development path for a country that can no longer compete in the market through the expansion of its manufacturing sector (Pirani, 2006). This has been described as a model based on 'many symbolic policies, lots of clever syncretism, very clear and illustrative messages, a variety of themes, a careful management of emotions and an awareness of the role that ceremonies can play in recreating social cohesion and the community dimension' (Ceccarelli, 2005, p15). In the words of Walter Veltroni, Mayor of Rome from 2001 until early 2008 (Veltroni, 2006, p144):

> *There is no real development without social quality. [...] If there is a Roman model, if many people today are talking in those terms about our experience, it is because everything we do aims to keep together economic growth and social cohesion and because at the foundation of every choice we make there is always a way of working, of collaborating, of 'concerting', of proceeding together: the*

> *municipality, the city council, and, with them, the business world, the trade associations, the social forces and the various actors of civil society.*

If, Veltroni states, development is about 'increasing social cohesion', then the primary obligation institutions have is that of 'not letting the distrust take over, not letting entire social classes lose heart and give up'. As the Mayor explains (Veltroni, 2006, p147):

> *It is as if there was a centrifugal force, an extremely dangerous one, which has the potential to detach entire sectors of the society and to draw them into a context of uneasiness and marginalization. A fundamental task that a government, an administration, has is to oppose to this force, to contrast it by pushing towards the inside: towards inclusion.*

In this model of 'a welfare state based on subsidiarity', the school meals system can be considered as an example of the 'capacity of keeping together innovation and competitiveness with citizens' lives, with an attention to the opportunities available for the talented and the needy alike' (Veltroni, 2006, p4). As stated by Rome's Councillor for Education, school meals are seen as an integral part of the wider role of the school as 'a place that promotes cultural, healthy and solidarity values that are important for the welfare of the entire community'.

By adopting a dynamic, inclusive and integrated approach, Roman authorities are designing a quality-based food system in which the economic relationships between producers and consumers are socially and environmentally embedded. This is not necessarily and exclusively about promoting local food in schools. Indeed, as we have described above, only recently has food re-localization become a priority in Rome's procurement policy. For years, the city attempted to meet the goals of sustainability by procuring Fair Trade, organic and certified products.

Specifically, through its Fair Trade sourcing policy, Rome is promoting more equitable forms of economic development by contributing to redressing unequal trans-national labour relations. At the same time, labour issues are also emphasized at home through measures that specify the number of staff that contracted companies must employ to guarantee a quality service and that reward them for employing additional staff. Democracy and social justice also play a significant role in the Roman quality revolution. By making the meals affordable and, simultaneously, diverse in their composition, Rome is providing access to school meals to all consumers, regardless of their socio-economic or ethnic background. Furthermore, as mentioned above, the city's procurement policy is currently bringing into the system charity associations and social cooperatives to assist the most vulnerable segments of the city's population.

As for the goal of environmental integration, over the course of its quality revolution, Rome has encouraged suppliers to shorten the supply chain, to source foods that have been produced through low-impact methods, to reduce noise pollution in the school canteens, to acquire environmental certificates and to put in place recycling schemes.

Rome's quality revolution is now extending beyond schools. In March 2007, city authorities, in partnership with the Region, decided to invest €84,000 to pilot the introduction of automatic machines selling 'healthy' snacks (seasonal and organic fruit and regional PDO cheeses) in 20 high schools, public offices and post offices located in Rome. As the Regional Councillor for education stated, 'this is a concrete answer to our citizens' demand for health. [...] At the same time, we are stimulating an economic growth that is compatible with the environment, health and nature' (Brera, 2007, piii). In accordance with the wider social objectives of Rome's quality revolution, the Region has decided that a portion of the profit accruing from the sale of fruit and vegetables will be destined for a farm created to feed children from Congo.

It is still too early to assess the concrete impacts of Rome's school food revolution in achieving the objectives of sustainable development. However, this case study highlights, at the very least, the power of the public sector in mobilizing support for sustainable development all across the food system. By complementing market interventions with food education and with the development of an ongoing dialogue with actors at both ends of the food chain, the emerging Green State in Rome is creating a shared vision for a sustainable future of economic development, democracy and environmental integrity.

5

A Sustainable World City? School Food Reform in London

> *London's extraordinary social and cultural diversity is reflected in over 60 different cuisines provided in over 12,000 restaurants – more than half the nation's total. However, there are also significant challenges. Obesity and diet-related illnesses account for a huge number of premature deaths in London, with many on low incomes suffering disproportionately. In many parts of London, people struggle to access affordable, nutritious food. Many of those involved in the food system are barely benefiting from it economically and the environmental impact of the food system is considerable. There are many features of London's food system that we need to improve if we are to meet my vision of a sustainable world city.* (LDA, 2006a, p3)

One of the conceits of world cities is the notion that, in an era of globalization, they have more in common with each other than with their own countries, as though they had decoupled themselves from their respective nation-states and formed a new, globally exclusive network of *city*-states. This notion is most prevalent in the upper echelons of the international business community, where hyper-mobile executives tend to live, work and play in much the same way whatever the city. This is especially true in London or New York, the world cities that have the greatest cultural affinity with one another. Indeed, such is the affinity between the two premier cities of the English-speaking world that the business community already refers to them by their composite name: *NyLon* (Gapper, 2007).

The London Plan, the Mayor's spatial development strategy, strongly subscribes to the world city view:

> *London is a world city and acts as one of a very small number of command and control centres in the increasingly interactive network of transactions across the world economy. World cities have very distinctive needs. Although separated by thousands of miles, they are intimately linked as a virtual global entity by the transactions of markets and communication systems. To reflect*

> *these links, the Mayor has begun to develop collaborative relationships with other major world cities.* (Mayor of London, 2006, p15)

This city-centric view is not necessarily untrue, but it is partial. World cities do have a lot in common with each other, particularly with respect to business practice and popular culture. Nevertheless, they are still intertwined with their domestic nation-states in social, economic, political and ecological terms. In a whole series of perhaps prosaic, but nonetheless important, ways, world cities are inordinately dependent on countless 'ordinary' activities, without which the daily rhythms of everyday life would quickly grind to a halt. They depend, for example, on the strong flows of commuters from the city's surrounding hinterland, which highlights the economic significance of the *city-region*, rather than the city per se. They also depend on the barely visible army of low-paid workers within the city – refuse collectors, bus drivers, nurses, waitresses and school cooks, among many others.

School food is a very instructive way to illustrate the schizophrenic status of London as both an *ordinary* city, which is part and parcel of its country, as well as a *world* city, which has unique features on account of its special status. The ordinary character of London is illustrated by the fact that its world city status has not saved it from the low quality of British school food; in this respect, London is very much part of its national culture. On the other hand, the combination of cultural diversity and urban scale renders the task of school food reform in this world city much more difficult than in the rural towns that we will examine in the next chapter.

To fully understand school food reform in London, we need to situate the campaign within a wider, multilayered context, where a dynamic interplay of national and capital city factors led to the dramatic re-emergence of school food as a political issue. To address these issues, the chapter is structured in three main sections. First, we examine the fall and rise of school food as a political issue in the UK. Second, we turn from the national to the city focus to examine the Mayor's strategy to transform London into a 'sustainable world city', a strategy that places a great emphasis on food and responsible public procurement. Finally, we explore the local experience of Greenwich, which was one of London's leading boroughs in terms of school food reform even before it was visited by a celebrity TV chef.

Three Eras of Welfare Reform: The Fall and Rise of School Food in the UK

The history of school food reform in the UK could be written in two radically different ways: either as a history of *welfare*, where the animating principle was the health and wellbeing of children, or as a history of *warfare*, where the motive was to furnish the armed services with a ready supply of well-fed and able-bodied working class recruits. In reality, even if welfare and warfare considerations appealed to radically different social and political constituencies, they both help to explain the origins and evolution of school food policy in Britain.

The welfare era of collective provision

Given the low prominence of school food as a political issue in the 1980s and 1990s, one might never think that it was once considered to be one of the pillars of the British welfare state. Social policy historians tend to locate the origins of school food provision in the 1880s, when the birth of compulsory education exposed the problem of undernourished children and their inability to learn effectively. Warfare became as important an influence as welfare when it was discovered that the poor physical condition of recruits during the Boer War was impairing the war effort. As a result, a Royal Commission on Physical Deterioration was set up; its report led to the Education (Provision of Meals) Act of 1906, which gave all Local Education Authorities (LEAs) the power to provide meals free for children without the means to pay for them and at a charge of no more than cost for other children (Passmore and Harris, 2004).

If the origins of the welfare era can be traced back to the 1880s, it was the Education Act of 1944 that really codified the values of this era of collective provision. Among other things, the 1944 Act laid a duty on all LEAs to provide school meals and milk in primary and secondary schools; it specified that the price of meals could not exceed the cost of the food; and it established that the school lunch had to be suitable as the main meal of the day and had to meet the nutritional standards that were first introduced in 1941, at the height of another war effort (Sharp, 1992).

Significantly, one of the original intentions of the welfare era was never realized. A parental guide to the 1944 Education Act outlined a radical perspective: 'when the School Meals Service is fully developed, school meals will be provided free of charge as part of the educational system' (Gustafsson, 2002). In this crucial respect, one of the key commitments of the welfare era was quietly forgotten by the post-war Labour Government, which otherwise did so much to lay the foundations of the welfare state. Whatever the limitations of the welfare era of collective provision, the fact that it would appear in retrospect as a 'golden era' speaks volumes for what followed.

The neo-liberal era of choice

Although the neo-liberal era was largely driven by a desire to reduce public expenditure, this was not the only factor at work. The school meal service had to adjust to the new consumer culture of the 1970s, when children were beginning to reject the traditional fare of collective provision. Changing 'consumer' behaviour in school canteens was marshalled as evidence to justify a radically different school food policy that chimed with old themes in Conservative ideology (less public expenditure and more private choice) and was pursued with unprecedented fervour by the Thatcher Governments.

The neo-liberal era was heralded by, and embodied in, two radically new policies. The first was the 1980 Education Act, which transformed the school meal service from a compulsory national subsidized service for all children to a discretionary local service. In all, the 1980 Act introduced four fundamental revisions: it removed the obligation on LEAs to provide school lunches, except for children entitled to free school meals; it removed the obligation for meals to be sold at a fixed price; it eliminated the requirement for lunches to meet nutritional standards; and it abolished the entitlement to free school milk (Passmore and Harris, 2004). The historical significance of the 1980 Education Act lies in the fact that, by design rather than default, it abolished the gains of the 1944 Act and jettisoned the values of the welfare era.

The Conservative education minister, Mark Carlisle, identified three reasons why school meals had to be reformed:

1 to make savings in public expenditure and to establish the principle of 'sound economics' for the parent, the taxpayer and the child;
2 to ensure that the burden of education expenditure cuts fell on the school meal service and not the education service itself; and
3 to give parents and children more freedom of choice.

Even though Parliament eventually approved the proposals, the minister was told that the service, apart from serving a very important nutritional purpose, also had a cultural value because it taught children how to eat a meal in the company of others (Hansard, 1979).

Although it was a Conservative Government that rolled back the 1944 Education Act, the previous Labour Government had already begun to cut the school meal service in the 1970s and, had it been returned to office in 1979, was planning to cut the service even further. We need to remember, as Tim Lang, the first professor of food policy in the UK, wrote, 'how poor Labour's record on school meals was, and how it was only embarrassed into raising the free meals level after Cabinet leaks to *New Society* magazine' (Lang, 1981).[1] One can assume that a neo-liberal government might have little sympathy for a key welfare measure like school meals, but it is more difficult to understand why the Labour Party felt unable or unwilling to mount a more vigorous defence of the service.

The second legislative vehicle for the neo-liberal era was the 1988 Local

Government Act, which introduced compulsory competitive tendering (CCT) into public sector catering. Under the CCT regime, local authorities were required to submit their school meal service to outside competition. As bidders felt obliged to offer the lowest price, CCT introduced a dramatic reduction in costs, which induced major changes in the school meal service. These include a less skilled workforce, a loss of kitchens (as a processed-food culture took over) and a service ethos widely deemed to be inimical to healthy eating. Of all the changes wrought by CCT, the most important was the debasement of the food itself, which was described by one prominent school cook as 'cheap processed muck' (Orrey, 2003).

Taken together, these twin legislative changes triggered a genuine revolution in the way the school meal service was designed and delivered, especially in secondary schools:

> *Most caterers opted for a cash cafeteria system, resulting in food being individually priced and the pupils having free choice. Children would spend as much, or as little, as they wanted, and there was no method of controlling what pupils ate. The school lunch service was very consumer-led and if a food sold well and was profitable, it was provided. If it did not sell, or was not profitable, it was not provided. Between 1980 and 1998 this strategy led to the current limited range of foods available in most secondary schools.* (Passmore and Harris, 2004)

From today's perspective, the neo-liberal era of school food policy appears to be responsible for a monstrously myopic mistake. In its desire to make short-term public expenditure savings, the Conservative Government actually contributed to the problem of unhealthy eating, which now costs the public purse many times what was saved by trimming the school meals budget. But before we are tempted to dismiss the neo-liberal experiment as a misguided historical curiosity, it is worth remembering that some of the values on which it was based – 'cheap food' and 'choice' – continue to resonate among many consumers and politicians, a paradox that we will address in Chapter 8.

The (emerging) ecological era: Sustainable provision and controlled choice

The fact that a radically new school food policy did not appear in England until 2005, eight years after New Labour came to power, illustrates the fallacy of thinking that regulatory eras change when governments change. In the public mind, the revolution in British school food policy is associated with *Jamie's School Dinners*, a very popular TV series featuring Jamie Oliver, a celebrity chef, who worked as a 'dinner lady' in London to expose the problems of the British school food system. His Feed Me Better campaign, which fired the public's imagination, is credited with changing government policy almost overnight.[2] However, the real origins of the British revolution lie not in a TV series set in London, but in Scotland, where, as we will discuss in Chapter 6, an expert panel produced *Hungry for Success*, a truly seminal report that called for a radically new school meal service (Scottish Executive, 2002). In particular, the report:

- promoted a whole-school approach to school meal reform, to ensure that the message of the classroom was echoed in the dining room;
- called for better quality food to be served in schools, supported by new nutrient-based standards; and
- suggested that the school meal service was closer to a health service than a commercial service.

The ripple effect of this Scottish social policy innovation stimulated the campaign for school food reform in England and Wales, though these countries did not produce their own equivalent reports until 2005 and 2006 respectively. Because it appeared so much later than *Hungry for Success*, perhaps, the English report, *Turning the Tables*, went furthest in embracing an 'ecological approach' to school food reform. Far from being purely concerned with the environment, the ecological approach is predicated on one of the core principles of sustainable development: the need to render visible the costs and connections that have been externalized (and rendered invisible) by conventional cost–benefit analysis, much of which is based on a rather desiccated metric of profit and loss. The report contained 35 major recommendations as to how and why England should implement a radically new kind of school food system. Echoing many of the recommendations made in Scotland, it signalled an even more holistic approach by including the food *procurement* process; as it stated, this should be 'consistent with sustainable development principles and schools and caterers should look to local farmers and suppliers for their produce where possible' (School Meals Review Panel, 2005).

As well as calling for more resources, new skills and higher food standards, the English report also challenged one of Tony Blair's most cherished ideological concepts: choice. In fact, *Turning the Tables* strongly endorsed the principle of 'choice control', which, it argued, 'has been shown to be effective not only for school lunches, but also in promoting healthier eating from other food outlets within schools'. Most of these recommendations were accepted by the government when it made its historic announcement, on 19 May 2006, that new standards would come into force in September 2006 to ensure that:

- school lunches are free from low-quality meat products, fizzy drinks, crisps, chocolate and other confectionary;
- high-quality meat, poultry or oily fish is available on a regular basis;
- pupils are served a minimum of two portions of fruit and vegetables with every meal;
- deep-fried items are restricted to no more than two portions per week;
- schools and vending providers promote sales of healthy snacks and drinks such as water, milk and fruit juices; and
- schools will raise the bar even higher when more stringent nutrient-based standards – stipulating essential nutrients, vitamins and minerals – are introduced in primary schools by September 2008 and in secondary schools by September 2009 (Department for Education and Skills, 2006).

The scope and limits of the ecological era

The euphoria that greeted the new school food standards in England could prove to be short lived, however. Translating the rhetoric into reality is a long-term undertaking and it cannot be achieved by administrative fiat from central government. At least three conditions have to be met before the promise of the ecological era can be delivered in practice to schools throughout the country.

First and foremost, there is the question of *extra resources*. The government announced £220 million over three years to help fund the transition to a healthier school food service in England.[3] This was designed to help schools make good the years of underinvestment in food ingredients, kitchens and training provision. Although it will certainly help to get the reform process under way, it may not be enough to put it on a sustainable footing. In fact, the new school food service is being launched at a time when many local school catering services are in a very fragile financial condition. Following the Jamie Oliver TV series, and in part caused by it, the take-up of school meals in the UK went down, placing school caterers in the unenviable position of dealing with an unsustainable process of higher costs and lower take-up.

Second, *new skill sets* are needed to implement the reforms. Caterers and cooks need to be equipped with healthy cooking skills. School governors and teachers also have to be mobilized because the ecological era is predicated on the whole-school approach, where the healthy eating message permeates the canteen, the vending machine and the classroom. Local authority procurement officers will need to acquire the competence and confidence to design and deliver tenders that allow local food to be the norm, rather than the exception. Finally, farmers and producers also need to acquire the skills to provide schools with fresh, locally produced food wherever possible. From farm to fork, then, new skills will have to be developed throughout the food chain.

And third, greater *social participation* is required if the reform is to be sustained. While the welfare and neo-liberal eras designed their policies for children, the ecological era will have to design its policies *with* children and their parents. A more deliberative and democratic system of governance will enable children to become active agents in their own transformation, rather than the passive objects they have been in the past. Children and their parents need to be more actively engaged in discussing the transition to healthy eating. As the Local Authorities Caterers' Association pointed out, the number of school meals fell by 10 per cent (equivalent to some 71 million meals) in the year following the Jamie Oliver TV series. As we will discuss in Chapter 8, children's taste cannot be transformed overnight; hence healthy eating needs to be understood as a socially negotiated process, rather than a technically conceived and discrete event.

Finance, skills and governance are certainly the chief concerns, but they are not the only ones. The Health Education Trust, for example, lamented the fact that the government refused to increase the eligibility for free school meals. This is a major worry for health campaigners because higher school food prices could further reduce the take-up rate of the midday meal, which is still the only hot meal of the day for many children.[4] If take-up rates fail to improve in future, the next phase of reform could unfold along the lines of 'free school meals for all', a policy that is demanded in Scotland and that was actually put into practice for a short time in the English city of Hull, where the take-up rate improved until the experiment was prematurely discontinued.[5]

Another problem that concerns school food reformers is the slow death of what used to be called the 'lunch hour'. More and more schools in the UK are trying to condense lunch into as little as 30 minutes, from a current average of 45 minutes, rendering it impossible for children to enjoy their food in a pleasant eating environment. Unless more time is found, it will be almost impossible to deliver the new school food standards.

If these problems can be resolved, the ecological era of school food reform will probably become the most significant era in social policy since the 1944 Education Act, the historic act that ushered in the welfare era of collective provision. Whereas the welfare era was a centralized system, the ecological era is a decentralized system, where the school food service is managed not just by local authorities, but, in some areas, by the schools themselves. Since such a devolved system makes it difficult to generalize from one area to another, we need to turn now to the world of local government to appreciate the nuances of school food reform.

Healthy Schools in London: A Food Strategy for a Sustainable World City

At the behest of its directly elected former Mayor, Ken Livingstone, London has embarked on an audacious project: to become a *sustainable world city*. Although some green campaigners might think that this term is an oxymoron, given London's enormous ecological footprint, the Mayor's strategy merits serious attention because it has the potential to make a genuine difference, not just for London but for the way we think about cities in general. In particular, the Sustainable World City (SWC) strategy helps to explore the meanings attached to the 'Green State' at a metropolitan level and highlights the need to strike a more judicious balance between globalization and localization. What is most interesting about the SWC strategy from our standpoint is the fact that it allots a central place to *healthy school food*, which is one of the priority actions of the Mayor's food strategy.

A good example of the interplay between devolution and development, the SWC strategy would probably not have emerged without the recently fashioned office of the Greater London Authority (GLA). Created in 2000, the GLA restored citywide strategic government to London after it had been abolished by the Thatcher Government in 1986, leaving the capital in the astonishing situation of having no metropolitan government. For a fast-growing world city of 7.5 million people, strategic citywide government was thought to be essential on both equity and efficiency grounds, and Tony Blair's New Labour Government established the GLA as part of its devolution programme (Morgan, 2007a).

Despite its status as a world city, the power of London's metropolitan government is more apparent than real because, in sharp contrast to New York, for example, it is highly circumscribed with respect to its powers and resources. The main problem is that the Mayor, along with the rest of the GLA, is actually squeezed between central government and local government (which consists of 32 separate boroughs and the City of London). As the leading expert on London governance puts it, the Mayor can at best try to use the legitimacy of his mandate to encourage other institutions to act by deploying his 'powers of patronage, persuasion and publicity – but these are no substitute for real fiscal and administrative authority' (Travers, 2004, p185).

To realize his SWC vision, Livingstone launched a series of imaginative strategies in transport and mobility, affordable housing, health, spatial planning, food and procurement. Without a doubt, the most visible achievement of his two terms in office was the successful implementation of the congestion charge, designed to reduce traffic volumes in the central zone of the city.[6]

Not surprisingly, in the light of what we discussed in Chapter 2, perhaps the least visible of all the former Mayor's policies for sustainability was public procurement. In 2006, the Mayor adopted a new sustainable procurement policy for all the public bodies for which he is responsible (known as the GLA Group), which collectively spend nearly £4 billion a year on goods and services. In the following year this policy was renamed *responsible procurement*, because Livingstone feared that the original name was overly associated in the public mind with environmental issues at the expense of social issues. As it operates in London, responsible procurement has seven distinct strands:

1 encouraging a diverse base of suppliers;
2 promoting fair employment practice;
3 promoting workforce welfare (London's Living Wage);
4 meeting strategic labour needs and enabling training opportunities;
5 providing community benefits;
6 practising ethical sourcing; and
7 promoting greater environmental stability.

'As a world class-city', Livingstone said, 'it is only right that we lead the way on making sure London buys its goods and services responsibly, and encourages other buyers and suppliers to follow, so that all communities fully benefit from our capital's prosperity' (Mayor of London, 2008b). The Mayor believes that responsible procurement will deliver significant benefits for London (listed in Box 5.1).

Among the first sectors to feel the effect of the new policy were catering and cleaning. To promote a fairer deal for these historically low-paid workers, the GLA was the first organization to introduce living wage provisions into its catering and cleaning contracts to ensure that no employees were paid below the living wage and that other employment privileges were not reduced as a result of paying a living wage. In addition to these social innovations, for the first time in history a preference for 'local food' was specified by City Hall on environmental grounds (Livingstone, 2006).

If procurement scores low on the public visibility index, the London Food Strategy (LFS) probably scores higher than any other GLA initiative, other than the pioneering congestion charge. Early in his second term, Livingstone created the London Food Board to develop a food strategy for the capital and appointed Jenny Jones, a Green Party member of the London Assembly, to chair it. Following widespread consultation, the strategy – 'Healthy and Sustainable Food for London' – was officially launched in May 2006.

Though it is clearly more aspirational than operational, the LFS is a significant component of the former Mayor's SWC vision. Among other things, it explores the significance of food in and for the capital; it sets out a

Box 5.1 *The benefits of responsible procurement in London*

Environmental benefits

- reduction in harmful emissions and waste generation
- improved air and water quality
- reduced use of natural resources

Social benefits

- improvements in working conditions – labour standards, health and safety
- assistance to disadvantaged groups in society
- improved and maintained standard of living

Economic benefits

- contribution to the modernization and international competitiveness of local industry
- improved efficiency in the public sector
- improved efficiency and transparency of procurement procedures/structures
- financial savings
- some sustainable products save money, whilst others may initially cost more, but save over the full life cycle
- GLA group members should benefit from a reduction in combined contract prices

Other benefits

- fulfilment of the Government's commitment to put the environment at the heart of policymaking and help to withstand increased public scrutiny
- meeting international commitments, including the EU requirement to integrate environmental protection into all public policies, the Rio Declaration and the Kyoto Protocol
- stimulating the market for green technologies

Source: www.london.gov.uk/rp

vision for the future of London's food system; and it outlines some of the key actions required to realize this vision. The whole strategy is predicated on a very simple observation: that the 'food system' in the capital is out of step with 'the ambition that London should be a world-class, sustainable city' (LDA, 2006, p17).

World cities are famous for their spectacular contrasts and London is no exception. While it was recently crowned the 'gastronomic capital of the world' on account of its fecund restaurant culture, other features of its food system are nothing to celebrate. For example:

- A rising number of Londoners, particularly children, are becoming obese.

Since the strategy was published, it has emerged that childhood obesity is more prevalent in London than anywhere else in England and Wales, with 11.3 per cent of 4- and 5-year-olds and 20.8 per cent of 10- and 11-year-olds now suffering severe weight problems.

- In some parts of the capital, people struggle to access affordable, nutritious food.
- The safe preparation of food, in the home and in London's plethora of cafés and restaurants, remains a key concern.
- Many small or independent food enterprises struggle to survive and this threatens the diversity and resilience of the capital's food system.
- The environmental consequences of the way London's food is grown, processed, transported and disposed of are profound and extensive. Food is responsible for 41 per cent of London's ecological footprint (LDA, 2006, p17).

Spatial scale and social character both distinguish the food system in London from the rest of the country, furnishing some interesting nuances as to how we understand 'the local'. In spatial terms, the scale is so vast that conventional definitions of the local have to be revised when applied to the capital. As regards farmers' markets, for example, the National Farmers' Retail and Markets Association accreditation stipulates that producers must come within 30 miles of the market, but London has been allowed to operate a 100-mile radius.

The social character of the system also makes a difference. Demand for ethnically and culturally specific foods is much higher than elsewhere in Britain and it is also growing at a faster rate because the capital is the largest recipient of international migration.[7] The need to ensure that London's diverse communities have access to culturally appropriate food creates a unique set of threats and opportunities in the capital. On the one hand, it may set *cultural limits* to the extent to which 'local food' can meet the capital's needs; on the other, it creates *cultural scope* for London to have a benign influence on the prospects for food producers elsewhere in the world.

One of the most innovative features of the LFS is the fact that it was designed along two different dimensions: the functional and the political. The functional dimension covers the entire food system, stage by stage, from primary production through to consumption and waste disposal. The political dimension covers five food-related policy themes that resonate with the Mayor's SWC strategy: health, environment, economy, society/culture and security. The Mayor hoped to integrate these two dimensions to promote a more sustainable food system in London by 2016. His ten-year vision for eating and consumption proposes a capital city where:

- Awareness of health and quality issues is high everywhere, especially among vulnerable socioeconomic groups.
- Parents are encouraged to provide healthy nutrition to their children from the pre-natal stage onwards, and, irrespective of their cultural background, mothers are helped to breastfeed their babies.
- Across the diversity of cuisines and cultures, food is promoted in London and elsewhere; Londoners are encouraged to eat in convivial settings; and the opportunity to spend the time they personally need to enjoy food and the eating experience is widely available.
- Children, including those with special dietary needs, have access to a range of nutritious, affordable and appealing food and drink (LDA, 2006, p66).

Although the LFS is first and foremost a vision for a radically new kind of urban food system, its authors stress that everything depends on 'concerted action' – that is, a collaborative effort from each and every stakeholder. To focus this effort, the strategy identifies six *priority actions*:

1 ensuring commercial vibrancy;
2 securing consumer engagement;
3 levering the power of procurement;
4 developing regional links;
5 delivering healthy schools; and
6 reducing waste (LDA, 2006, p69).

Taken together, these actions cover many of the elements of a sustainable food system, and it is salutary to see that the power of the public plate has been recognized at last, not least because this could be a particularly effective mechanism through which the capital can fashion a more localized food system by reconnecting with its regional hinterland. As for the healthy schools priority (see excerpts in Box 5.2), the actions listed recognize that healthy school food requires a series of complementary investments, for example in updated catering skills, better catering facilities (kitchens and dining halls) and more robust political commitment.

Whether the LFS is assessed as a whole, or just in terms of its healthy schools priority, its prospects in terms of delivering the three goals of sustainable development are the same. Its success will depend mostly on two things: resources and governance.

If the available resources are limited to the official budget of the LDA, which has initially been allocated some £4 million over three years, we can safely say that this sum is far too modest to realize the Mayor's grand vision. The architects of the LFS are fully aware of this problem. As they stated:

Box 5.2 *Delivering healthy schools in London*

Schools have a fundamental role in the food system in London: they have the opportunity to provide pupils with healthy meals at least once a day; they can educate children about food, nutrition, healthy eating and the environment; they can equip children with the skills they need to make informed choices and prepare their own food; and they can equip children to educate and pass on knowledge to their parents and peers. More than any other group in London, children need, indeed are entitled to, strong guidance.

Focusing on all of these opportunities offers the scope for both immediate and longer-term health, behavioural and environmental benefits. This is not an easy win or short-term objective; there are indeed a number of significant barriers to overcome, including problems with catering skills, the lack of flexibility in some existing contracts with suppliers, appropriate cooking facilities and the level of funding overall. However, the potential benefits are such that London-wide action is required now. For this reason, the following key actions are proposed:

- Support the education system in increasing the amount of time spent on cooking and food education in schools, which may include work to revise the National Curriculum as well as specific support measures for schools and teachers.
- Research and promote the positive benefits of nutritious food for children, and work to secure the necessary funding and investment to secure those benefits.
- Continue to improve the nutritional quality of school meals and the number of pupils eating them, targeting areas such as training for catering staff, catering facilities, political will and overall budget allocations.
- Improve children's access to healthy, quality food outside of school meals by improving the provision of fresh fruit and access to fresh water in schools, supporting and piloting the introduction of green/healthy vending machines, and expanding school breakfast clubs.
- Increase the number of schools taking part in farm/city farm visits.

There is considerable momentum behind these issues – at both a national level and within London – that this strategy needs to capitalize and build on. For example, in London much good work has been done already in Croydon, Greenwich and Camden.

Source: LDA (2006), p95

The cost of improvements to London's food system cannot be met by the public sector alone. It will be vital to maximize the input and impact of the private sector, as well as voluntary organisations and, of course, individual consumers, on an equitable and enduring basis. (LDA, 2006, p103)

The same applies to governance, which will also require concerted action from a plethora of national, regional and local bodies. 'If the Strategy is to be effective', its authors said, 'it must engage a broad church to take forward its ideals with concrete actions, and this will require support across the entire food chain' (LDA, 2006, p101). However, this is easier said than done, because one of the inherent weaknesses of the LFS – perhaps its biggest weakness – is that the Mayor has little or no control over the activities that he is seeking to influence.

This disjunction between control and influence is especially evident in the school meal service, where control lies with the London boroughs, rather than with the Mayor. The governance of the capital's schools is a complex matter. Traditionally, the locus of responsibility for state schools in England lay with local authorities, but schools have become much more autonomous since 1990, when the advent of local management of schools transferred many functions to the board of governors of each school. In London, responsibility for schools is shared between 33 separate boroughs and the individual schools, making it virtually impossible to generalize from one area to another. This is especially true for the school meal service, which assumes a bewildering array of forms in and between boroughs. A study of school food procurement in London in 2005 found that just 12 boroughs used their own in-house catering facility – that is, their Direct Service Organization (DSO) – to supply school meals, which means that more than half of the 33 boroughs employed the catering services of a private contractor. Although good practice was found in both public and private sector catering provision, the study concluded that, where the council service is good, the DSO model offers the most scope for effecting change, because it is capable of reaching more schools (Sustain, 2005).

This smorgasbord system of provision suggests that, in at least some cases, the London boroughs themselves may have little or no *direct* influence over school food provision, making it much more difficult for them, either individually or collectively, to steer the system in a new direction. To understand what happens at the local level, therefore, one has to enter the world of the boroughs, which is the front line of school food reform in London.

The Front Line: School Food Reform in Greenwich

London's boroughs cover all manner of life in the capital, ranging from the glitzy lifestyles of central London to some of the poorest people in the country in the east. Despite being one of the poorest boroughs, Greenwich has been in the vanguard of school food reform in the capital. Although the borough assumed national prominence when it was selected by Jamie Oliver to illustrate the lamentable state of British school meals, as we will see, the celebrity factor accelerated, rather than initiated, school food reform in Greenwich.

Greenwich is an unusual borough in many ways. Whereas most London boroughs have outsourced school food provision to private catering companies, all but five of Greenwich's 88 schools are supplied by Greenwich Catering, the Council's in-house catering service. When it works well, DSO provision has many advantages, not least because it enables a council to furnish nutritious food on a borough-wide basis. This is especially important in Greenwich, which has one of the highest levels of deprivation and ethnic diversity in London.

Over a fifth of its 222,000 residents are under 15 years of age, a quarter of whom live in families receiving some form of income support. As eligibility for free school meals is deemed to be a robust index of poverty, it is worth noting that 38 per cent of all primary school pupils in Greenwich are eligible, which is more than double the English average of 17 per cent.[8] Despite these daunting levels of poverty, some 90 per cent of all schools in the borough will have acquired National Healthy School status by the end of 2008 (Greenwich Council, 2007).

The ethnic and cultural diversity is best conveyed by the staggering fact that more than 100 languages are currently spoken in the borough. Some 26 per cent of the population are non-white; of these, the largest groups are Black and Black British and Asian and British Asian. Significantly, the ethnic mix of the borough changes quickly, with West Africans, East Europeans and Iraqi refugees among the most recent arrivals. This raises continuous challenges for local public services and particularly for the school food service, which has to cater for rapidly changing dietary needs.

Social and cultural diversity also raises special health challenges. For example, the prevalence of obesity among black Caribbean women is 50 per cent higher than the average, and among Pakistani women it is 25 per cent higher than the average. Similarly, obesity in Asian children is almost four times more common than in white children. Life expectancy in Greenwich is much lower than the average for England, though there are significant variations within the borough.

Local politics helps to explain why Greenwich Council is committed to school food reform and why it decided to use its own in-house catering service, Greenwich Catering, as the vehicle to deliver it, rather than outsourcing the service to the private sector. With 36 of a total 51 seats, Labour is firmly in control of Greenwich Council, and to date the Party has been totally committed to providing healthy school meals throughout the borough. Councillor Chris Roberts, the Labour Leader of Greenwich Council, is clearly proud of what has been achieved:

> *Both Greenwich Council and our schools remain as committed as ever to healthy school meals, as part of our wider determination to improve the health of our young people. With more than three years of experience with the new menus, children, parents and staff now view healthy, freshly cooked food as the norm and don't expect to see anything different on their plates. The success of our school meals has been achieved through the tremendous dedication and commitment of staff and schools, and with great support from parents and children. We are extremely proud of what's been achieved in Greenwich, and of the impact that our success has had on school meals across the country.* (Roberts, 2008)

If political support has been an important ingredient in the recipe for school food reform in Greenwich, so too has the local catering service, Greenwich Catering, which survived the decimation of local authority catering in the Thatcher years. As Bobbie Bremerkamp, Operations Manager of Greenwich Catering, describes:

> *Within Greenwich, we had very strict control. We were very fortunate. We still kept our kitchens. We still kept our skills. We still produced one proper cooked meal a day. We made sure that was on the menu.*

Being able to retain cooking skills and kitchen facilities has been vital to Greenwich's ability to undertake a major reform programme in a remarkably short space of time. Equally important was the fact that, even during the most difficult neo-liberal years, Greenwich Catering successfully retained the contract for all Greenwich's primary schools and, though it initially lost the majority of its 13 secondary schools, has since won back the business of 11 of them. Fortunate as it was, however, Greenwich Catering did not emerge unscathed from the neo-liberal era. Ms Bremerkamp recalls:

> *What happened was, after every four or five years, you had to keep tendering. So every time it was tendered, the prices were driven down. So the food items were driven down. And with that the quality went down. And it got to the point that you were embarrassed by some of the things that were coming in.*

Deconstructing celebrity: The scope and limits of the 'Jamie Oliver effect' in Greenwich

The celebrity embrace is invariably a mixed blessing, as it proved in Greenwich, when Jamie Oliver descended on the borough with a TV crew to expose the woefully inadequate British school meals system. The chef taught the local dinner ladies many things – cooking with fresh ingredients, introducing new recipes, designing new menus and working with local suppliers, for example – but he also learnt a lot from them. Among other things, Oliver was forced to come to terms with the front-line pressures of the school food service in a poor community – in particular, the pressure of phenomenally tight budgets, the pressure of having to serve 5600 children in 45 minutes and the pressure of trying to wean children off their apparent addiction to junk food. The world of exclusive restaurants has its own pressures, of course, but they are nothing like as challenging as the pressures Oliver confronted when he adopted the persona of a 'dinner lady' in Kidbrooke, a school that draws its pupils from one of the most deprived areas in Greenwich. Exposed to the real world of mass catering, Oliver was forced to revise his menus to ensure that what was offered could be delivered within the constraints of cost and time.[9]

Celebrity status also enabled the new 'dinner lady' to secure local political support almost overnight. One of the first things Oliver did was to invite the local education establishment – head teachers of the schools and the Leader of the Council – to a dinner with a difference at his restaurant in central London. Cooked by the dinner ladies of Greenwich, the event was designed to highlight a quality meal that could be served in the borough's schools. Greenwich's Director of Culture and Community Services, Gurmel Singh-Kandola, described the dramatic reaction from his guests: 'And then and there, that night, all the head teachers signed up to it. As soon as they signed up to it, Bobbie had the clear mandate to begin a roll-out programme.' Here we have another illustration of the power of celebrity, which in this case helped to secure the 'buy-in' of the local political and educational establishments – something that might have taken months, or even years, had it been done through the normal bureaucratic channels.

If politicians and head teachers were easily won over to the cause of healthy eating, Jamie Oliver had a much harder time winning the hearts and minds of children and their parents. According to Bobbie Bremerkamp, the chef was chastened and unnerved by the initial reaction to his menu, most of which was left uneaten and had to be discarded as waste. Ms Bremerkamp also remembers the parents' reaction:

> *Well, the parents really retaliated against him. You know, you had parents coming in – 'My daughter won't eat your food – it's rubbish.' 'What does your daughter eat?' 'She'll only eat burgers.' And that was sort of what they were telling us. They wanted burgers put back on the menu. These children were free school meals children, but the parents voted with their feet and sent them in with packed lunches.*

Reflecting on this chastening experience, Oliver freely concedes that his short career as a 'dinner lady' was one of the hardest things he has ever done. As he states on his website, the only way he could persuade the Kidbrooke kids to eat the healthier food was by ensuring that there was no alternative (see Box 5.3).

Box 5.3 *How the Feed Me Better campaign happened*

When I signed up as a dinner lady to make school dinners at Kidbrooke school in Greenwich, I wanted to show people what rubbish their kids were getting fed at school and how little government was spending. Basically, I wanted to get rid of the junk. I had to prove that, for the same price as a bag of crisps, just 37p, I could produce a properly cooked, nutritious meal at lunchtime.

I had no idea that Jamie's School Dinners was going to be the start of such a massive campaign and that it would get such enormous support – from parents, teachers and kids all over the world.

The Jamie's School Dinners TV programme

At Kidbrooke, the kids were eating a quarter of a ton of chips every week. The food budget was 37p per meal and Nora and her team of dinner ladies had become totally unmotivated by the food they served.

I needed a set of menus that we could serve to the 15,000 schoolchildren across Greenwich, for the same price, to set an example for the whole country and show the Government it could be done. With help from the army, we got all 50 head dinner ladies together and trained them to peel and chop veg, and to make their own fresh meals again. Afterwards, each school had a chef working behind the scenes for a week to help them get up and running with the new menus. To get the kids to accept the new food, I had to [make unavailable] all of the junk food so there was no alternative. It was the one of the hardest things I've ever done, but it worked. A year later, Nora's still serving the same food, the kids love it and they have a portion of fruit and vegetables or salad with every meal.

Feed Me Better

The Feed Me Better campaign was fantastic. I needed loads of support to go to the Government and get them to change, so I wrote a manifesto asking for five things:

1. guarantee that children receive a proper, nutritionally balanced meal on their plates;
2. introduce nutritional standards and ban junk food from school meals;
3. invest in dinner ladies: give them better kitchens, more hours and loads of support and training to get them cooking again;
4. teach kids about food and get cookery back on the curriculum; and
5. commit long-term funding to improve school food.

Suddenly, School Dinners was making the front pages of all the papers. We set up a web petition and hoped to get 10,000 people to sign up. After a week we already had 25,000 signatures.

Source: www.jamieoliver.com/media/jo_sd_history.pdf?phpMyAdmin=06af156b76166043e2

A year after the new menus were rolled out across the borough, the children seem to have taken to the healthier meals, though it is worth highlighting the fact that take-up rates have not increased dramatically, proving that children's tastes are not taps that can be turned on and off overnight. On the whole, however, children are eating more nutritious meals; as shown in Figure 5.1, Greenwich Catering uses only fresh fruits, vegetables and meat in their menus.

Sample School Menu

All dishes are homemade – All vegetables and salads are fresh and prepared at the school
The vegetables are a guide. All meat is fresh and UK sourced.

* = Meat Option ** = Fish Option V = Vegetarian Option

	MONDAY	*TUESDAY*	*WEDNESDAY*	*THURSDAY*	*FRIDAY*
WEEK 1 Bread & Salad Bar Everyday	* Proper Sausages Creamy Mash Peas & Sweetcorn ■ Mexican Bean Wrap (v) Cheesy Leek Pasta (v) Peas & Sweetcorn Salad ■ Vanilla Sponge & Custard	* Chicken & Mushroom Casserole * Chilli Con Carne Savoury Rice & Salad ■ Vegetable Chow Mein (v) Salad ■ Fruit Crumble & Custard	* Roast Beef Roast Potatoes, Green Beans & Gravy ■ Mushroom & Lentil Bake (v) Roast Potatoes & Green Beans ■ ** Tuna Jacket Potato Green Beans ■ Fresh Fruit Platter & Custard	* Lamb & Vegetable Pie Veggie Mince Pie (v) ** Creamy Coconut Fish New Potatoes Broccoli ■ Creamed Rice Pudding	* BBQ Chicken Cheese Flan (v) Jacket Wedges Salad ■ * Cottage Pie Seasonal Vegetable ■ Fresh Fruit & Custard
WEEK 2 Bread & Salad Bar Everyday	* Home Made Sausage Rolls Country Diced Potatoes, Baked Beans ■ Vegetable Chilli (v) Country Diced Potatoes ■ * Italian Pasta Bake Salad ■ Fruit Sponge & Custard	* Sweet & Sour Chicken ** Fish Korma Rice & Peas ■ Macaroni Cheese (v) Peas ■ Carrot Cake & Custard	* Chicken Tikka Masala * Balsamic Beef Vegetable Goulash (v) Rice Green Beans ■ Apple Pie Custard	* Moussaka ** Mediterranean Fish Carrot & Cheese Slice (v) French Bread Sweetcorn ■ Fresh Fruit or Yoghurt	* Lemon Chicken Veg Mince Cottage Pie (v) New Potatoes Salad ■ Vegetarian Cannelloni (v) Salad ■ Fresh Fruit Salad & Custard
WEEK 3 Bread & Salad Bar Everyday	* Proper Sausages New Potatoes, Peas & Gravy ■ * Egg & Bacon Quiche New Potatoes & Peas ■ Leek & Lentil Pie (v) Peas, Salad ■ Shortbread Biscuit & Custard	* Roast Turkey Roast Potatoes, Cabbage & Gravy ■ Cauliflower Cheese (v) Roast Potatoes & Cabbage ■ * Chilli Beef Fajitas Salad ■ Fruit Sponge & Custard	** Herb Crusted Fish * Beef Goulash Creamy Mash Broccoli ■ Bean & Cauliflower Bake (v) Salad ■ Peach Crumble & Custard	* Chicken Curry Rice Salad ■ Vegetarian Lasagne (v) ** Tuna Pasta Bake Salad ■ Fresh Fruit or Yoghurt	* Shepherd's Pie Carrots & Peas ■ * Chicken with Noodles Cheese Jacket Potato (v) Salad ■ Lemon Sponge & Custard
WEEK 4 Bread & Salad Bar Everyday	Cheese & Tomato Pizza (v) Curried Vegetable Pasty (v) Oven Roasted Midis Salad ■ ** Tuna Arabiatta Salad ■ Rice Pudding	* Cajun Chicken * Lamb Curry Rice, Cucumber Raita & Salad ■ Cheese & Potato Hot Pot (v) Cucumber Raita & Salad ■ Apple Pie & Custard	* Lasagne Spicy Cheese Wraps (v) Salad ■ * Lemon Roast Chicken Country diced Potatoes, Cabbage & Carrots ■ Vanilla Sponge & Custard	* Spaghetti Bolognese Vegetarian Spaghetti Bolognese (v) ** Salmon Fish Cakes Garlic Bread Salad ■ Cherry Sponge & Custard	* Chicken & Mushroom Pie Egg & Cheese Flan (v) * Mediterranean Lamb New Potatoes Green Beans ■ Fresh Fruit or Yoghurt

For further information please contact: Operation Catering Manager on 020 8921 4621 or e-mail: cateringservices@greenwich.gov.uk

Greenwich Council

Figure 5.1 *Sample school menu in Greenwich*

Source: www.greenwich.gov.uk/Greenwich/Learning/SchoolsAndColleges/ImprovingOurSchools/HealthySchoolMeals/SampleSchoolMenus.htm

'Before Jamie Oliver', Ms Bremerkamp said, 'everything except fruit and veg was frozen, though years ago it had been fresh'. This is not the only example of déjà vu in Greenwich Catering – as Bremerkamp added, 'We've gone back to our existing cookery book that we used to have in ILEA days,[10] but we've updated that with spices and garlic, because things are more spicy now.'

Although *Jamie's School Dinners* is widely believed to have induced a fall in the national take-up rate for school meals, the changes associated with the celebrity chef in Greenwich are generally perceived to be positive. The question that is endlessly debated in the borough is whether any of these changes would have occurred without such celebrity intervention. Mr Singh-Kandola, a director close to the process, believes that the 'Jamie Oliver effect' propelled school food reform to the top of the political agenda in Greenwich Council, thereby circumventing the normal bureaucratic hurdles that impede radical change in any organization, especially local government:

> *If you want to make a change, it's not easy. With Jamie Oliver coming in, he became like the figurehead and, because of his kind of status, people were more willing to engage with it and that really helped, I have to say. The transformation which took place within the space of like six months, would have taken us years here. And, I can't emphasize enough, I've never seen a change driven so quickly and with so much willingness.*

A rather different interpretation, however, is offered by Claire Pritchard, who runs a food access project for the Greenwich Cooperative Development Agency. Accepting that Oliver raised the profile of the catering service, she argues that Bobbie and her team had already started the reform process, but they had no profile, no voice and no political support at the highest levels of Greenwich Council. As Claire Pritchard explained:

> *The truth is that people like Bobbie were totally capable of overhauling the school meal system without Jamie Oliver coming. They have the skills and they'd been asking for those things, but they just couldn't be heard. And that's an issue. The things they asked for, they didn't get until Jamie Oliver asked for them.*

The fact that school catering staff 'couldn't be heard' will resonate deeply among school caterers throughout the country, considering that often they have neither the profile nor the voice to make themselves seen and heard by service directors and political leaders in local government. Claire Pritchard makes another pertinent point when she says that, had it adopted a more 'joined-up' response, Greenwich Council could have met the challenge of school food reform without suffering such a hostile parental backlash to the healthy menus.

Greenwich had begun to develop an exemplary food and health strategy as early as 2003, when a partnership between the Council, the Primary Care Trust and the Greenwich Cooperative Development Agency was formed to improve nutrition in the borough. As a community development resource, the partnership could have been harnessed to mobilize the support of parents for the new school menus. In practice, however, it was ignored until the Council realized it was in a crisis with its poorest families. As Pritchard explained:

> *If you're doing community development, and you're really in touch with parents and children, then you can be really supporting [school food reform]. It is a really positive partnership now, it's just that it could have been easier for some of the people on the ground initially. Now we've got something very, very positive happening.*

In retrospect, it seems clear that Greenwich Council, while it eventually met the challenge of celebrity-induced change, could have managed this process better had it utilized its own soft infrastructure of community engagement.

Embedding the Gains: Fashioning a Sustainable School Food Service?

The real test of school food reform lies not so much in being able to attain a new quality standard but in having the capacity to maintain it when the glare of publicity has receded and when the political agenda has moved on. As we have seen, Greenwich was fortunate because it had not regressed, like many of the other boroughs, with respect to the terms and conditions of its catering staff, its stock of cooking skills and the infrastructure of kitchens, all of which were enhanced with new staff, training programmes and equipment upgrades. Six new quality managers were recruited for the transitional period and these were complemented by two permanent positions in finance and operations. Some extra funding was made available from central government to meet the transition to higher standards, but this amounted to just £137,000 for the whole of the borough. As Mr Singh-Kandola pointed out:

> *I don't want to be disrespectful, but in the scheme of things it had very little impact. So the majority of the money that we had to find was found by the Council and what we have invested is a very significant amount. In 2005–2006, we invested £629,000; in 2006–2007, £600,000.*

For poor, cash-strapped London boroughs, these are not trivial sums. With Greenwich Catering putting more emphasis on fresh fruit, vegetables and meat, Greenwich Council has to keep abreast of these changes in its food procurement tenders. Although Greenwich has a green procurement policy in place, the Council's officers admitted that 'it's not being pushed in any great way at the moment', highlighting that the pace of school food reform is manifestly uneven across the Council's departments. However, the new tender specification includes a copy of Greenwich Council's Local Food Policy, which explicitly states that 'the primary specifications for food purchased by the Council will relate to the nutritional and health aspects'. The procurement challenge for Greenwich Council lies not with its specifications, but in ensuring that they are taken seriously and acted upon.

For its part, Greenwich Catering is working hard to enhance its take-up rate, which is critical to the financial viability of the service. One of the innovations that it has introduced to retain student business is to cater for children who like to eat quickly and socialize during the lunch break. Bobbie Bremerkamp described the problem and the new solution:

> *In secondary schools, the numbers did fall off because, with the type of food we're serving, you need to sit down and eat with a knife and fork. And it takes too long, and they have to queue too long. Whereas before they got a burger and whatever and they were out. So what we've been doing in secondary schools is offering remote areas where we're serving paninis and a bowl of salad and a piece of fruit for one pound fifty, which is the price of a free school meal in the secondary school.*

These remote service areas are out in the playgrounds and the students can eat outside with their friends, much like they would at a fast food environment. The innovation, which resembles the approach adopted in New York City, has proved successful for students and caterers alike, earning much needed revenue for the secondary schools. A seemingly modest innovation, this is a good example of the critical variables – namely, cost, time and taste – that cooks and caterers have to juggle with every day of the school term – the very same variables that confounded Jamie Oliver.

The new menu varies from one end of the borough to the other, depending on ethnic mix and local tastes. As Bobbie Bremerkamp explained:

> *We do balsamic beef and it goes down a storm in one school, and in another they won't touch it, but put a lid of pastry on it, and you've got beef pie, and they eat it!*

Catering innovations are also occurring at the level of individual schools. In some cases the children participate in the design of dining areas, and one school has created its own vegetable garden to serve the school kitchen. The main limitation at the moment, other than finance, is that not every school has its own kitchen, though that remains the long-term goal of the healthy school food strategy in Greenwich.

Through the Greenwich Cooperative Development Agency and Sustain, the borough is also involved in a number of innovative food projects funded under the auspices of the LFS. These projects aim to redress key weaknesses in the capital's food system. One of them is based on the simple, but fundamentally important, idea of using school catering expertise and kitchens to teach children new cookery skills. Another project involves the creation of an integrated public catering training facility, based in Greenwich, to equip school cooks throughout London with healthy cooking skills. School food reformers in Greenwich are also exploring ways to aggregate public sector purchasing with schools, hospitals and care homes from other boroughs to enhance their procurement power in the food chain. These demand-side initiatives are complemented by plans to re-organize the supply side, especially the distribution infrastructure. Under the umbrella, once again, of the LFS, Greenwich is exploring ways to create an alternative East London hub for more sustainable regional food.

These initiatives in training, procurement and distribution are highly significant for both the borough and the city. For the borough they suggest that a ripple effect is beginning to emerge: what began as an apparently simple initiative to reform the school food service has exposed weak points elsewhere in the food system, highlighting the need for an integrated approach that can do justice to the interconnections between food, health and sustainable development.

For the city, recent initiatives suggest that some of the weaknesses identified in the LFS, like the disjunction between the Mayor's aspirations and his powers, may be more apparent than real. While it is certainly true that the Mayor has little or no *direct* control over the school food service in the boroughs, neither do the boroughs where schools have assumed full control of their own affairs. As we have seen, Greenwich has retained the loyalty of all of its primary schools and the majority of its secondary schools, which speaks volumes for the quality of service of Greenwich Catering. The joint projects between the borough and the city under the LFS also suggest that we should not confuse control with influence. The Mayor might not have any direct control over the school food service in Greenwich, but his LFS clearly resonates with what school food reformers in the borough want to achieve themselves. Neither the city nor the borough can achieve their goals acting alone, but they can achieve them by acting in concert. As is often the case with school food reform, progress tends to be secured less through command and control and more through partnership and persuasion.

6

Beyond the City: The Rural Revolution in School Food Provision

Our vision is of a partnership between children, school, family and the community in offering access to attractively presented food of an appropriate nutrient composition within schools and in developing a wider understanding of food, nutrition and healthy lifestyles which can inform children's choices and eating habits within and outwith school and throughout life. (Scottish Executive, 2002)

In the context of the school food revolution currently under way in the North of the world, some of the most surprising results have been achieved by local authorities located in rural areas. In Italy, as we saw in Chapter 4, small rural towns started practising sustainable food procurement long before the issue began to attract the attention of the Government or of procurement officers working in large cities like Rome. In the US, the Farm-to-School movement has achieved its best results mostly in rural areas of California, Massachusetts, Michigan, New Hampshire, North Carolina and other Southern states (Joshi et al, 2006), rather than in metropolitan areas of the country.

This trend is even stronger in the UK, where the rural counties of South Gloucestershire (England), Carmarthenshire (Wales) and East Ayrshire (Scotland) have played a pioneering role in the design and delivery of healthy school food that is locally sourced wherever possible. Far from being the backward areas they are deemed to be in some city-centric theories, these rural areas, which are often ignored by the metro-biased media, have taken the lead in making the connections between school food, personal wellbeing and the sustainable development of their communities. In this chapter, we will be telling the stories of school food reform in South Gloucestershire, Carmarthenshire and East Ayrshire to trace the origins of the British school food revolution. With their different emphases on social justice, community development and local sourcing, these three case studies speak volumes for the key challenges that rural areas committed to sustainable school meals are facing around the world.

'Not Simply Dinner Ladies': School Food Reform as a History of Labour in South Gloucestershire (England)

The county of South Gloucestershire, in the Southwest of England, differs quite significantly from the other rural areas that we will discuss in this chapter. Indeed, in contrast with Carmarthenshire and East Ayrshire, South Gloucestershire is characterized by a vibrant economy, which has experienced rapid expansion in the last 15 years. For example, the number of jobs in the county has increased from 91,000 in 1991 to 147,500 in 2007, when South Gloucestershire achieved the exceptionally high employment rate of 87 per cent, compared with a national average of 78.5 per cent (Godwin et al, 2007). With its 250,000 inhabitants and a catering division delivering 4 million meals a year to 125 schools, South Gloucestershire is also bigger than its Scottish and Welsh counterparts (see Table 6.1).

Table 6.1 *School meals in three rural areas*

	South Gloucestershire	Carmarthenshire	East Ayrshire
Population	250,000	180,000	120,000
Number of schools involved in the reform	125	150	26 (in 2007)
Price of a school meal	£1.45 (in 2007)	£1.65 (in 2007)	£1.52 (in 2007)
Take-up rate	54% (in 2006)	70% (in 2005)	44% (in 2006)

When it comes to school meals, however, there is nothing distinctive in the early history of the service in South Gloucestershire. Like all other counties in the UK, in the late 1990s South Gloucestershire had to face the devastating consequences of compulsory competitive tendering (CCT) that we discussed in Chapter 5. As the Head of Catering recalled, at the time the school meal service was 'floundering', with just 22 per cent of children eating at school (Knight, 2004, p31). Even more important, it was a service with no value and dignity:

> *The service had no value, and because it had no value, there was no money attached to it. [...] The benefits of proper service, proper food, proper interaction and social care, all those things were not seen. [...] There was no recognition above the core delivery of the food on the plate that it had any value whatsoever [...] to supporting a child's wellbeing or to the development and enrichment of learning.*

The consequences of years of commercialization of the service were catastrophic on a number of grounds. On the one hand, South Gloucestershire was facing major infrastructural problems: schools were increasingly getting rid of kitchens to utilize the rooms as storage space. At the same time, the equipment available to dinner ladies, as the Head of Catering described it, 'was very much old, out-of-date, [...] a very low priority [...] in the spectrum of educational provision'. On the other hand, CCT had 'eroded the terms and conditions of the staff':

> *The service had no dignity, and the people who were working in it had no dignity. [...] To take casuals and turn them into full-timers, you're having to pay them sick pay; you're having to pay them annual leave. If 70 per cent of the workforce is casuals or temps, what you are doing is [...] making the service cheap; it's got no value. [...] It wasn't driven for value, it was driven to save.*

South Gloucestershire was one of the first local authorities in the UK to understand that school food reform requires primarily a change in political culture. As the Head of Catering compellingly argued:

> *The CCT era brought 'the client' and 'the provider' – God help us all. And the client and the provider was the biggest stumbling block to getting integrity and value into the service in my opinion, because it created a two-tier level of delivery. [...] The contract was set up by the client; it left no leeway for the contractor to really develop it, innovate it, make it work.*

Indeed, South Gloucestershire's catering unit could count on the support of a council that has been committed to integrating the principles of sustainable development into its policies since 1996 (South Gloucestershire Department for Children and Young People, 2005, p7). Convinced that there was not much left to lose, and prepared 'to risk thinking out of the box', as the Head of Catering said, in 2001 South Gloucestershire Council agreed to entrust the service to an in-house team directly managed by the Department for Children and Young People.

Since then, South Gloucestershire has placed labour issues at the heart of its school food reform. In the vision of this local authority, economic and environmental sustainability cannot be achieved without a well-trained and highly motivated catering staff. To quote again the Head of Catering:

> *You have to have staff that understand the product. [...] We've got 125 sites; I cannot sit in every one making sure they're not wasting*

> *things. But you cannot manage the resources unless the staff own the service, because [otherwise] what difference does it make to them if they tip ten tonnes of beef into the bin? Why should they bother? [...] If they don't have an ownership, a value, an understanding, what are you trying to achieve?*

Under this approach, school food reform became primarily an issue of labour. As the Head of Procurement put it:

> *That's the key: self-respect and dignity. It's persuading people that they are not [...] simply dinner ladies. There is a value and a worth to the delivery.*

In 2002 and 2003, kitchen staff underwent an intensive training programme on nutrition and customer care. They were consulted in relation to work that needed to be done in the different schools to enhance the dining rooms and kitchens (redecoration and replacement of the furniture and/or of the kitchen equipment) (Knight, 2004, p32). And they were provided with new 'white' uniforms that they themselves chose, to signal, as the Head of Catering said, that:

> *This wasn't a fast food service. This was a service that needed cooks, needed chefs, needed innovation.*

South Gloucestershire's emphasis on social justice also manifested itself through other initiatives that were undertaken specifically to embed a 'culture of inclusiveness' (Knight, 2004, p35) in the school meal service. For instance, the Council developed a policy of non-identification of those entitled to free school meals. In primary schools, there are no lunch tickets issued to children. Meals are paid cash to a school administrator, with class teachers recording which pupils are staying for lunch. Through this system, students merely line up at the point of service to receive their lunches and there is no differentiation between paying and non-paying children (Knight, 2004, p34). In addition, South Gloucestershire has introduced 23 breakfast services within schools located in needy areas at a very affordable price (for example, egg on toast is sold at 15 pence). School meals cost parents the reasonable price of £1.45 – a figure, the Head of Catering said, which has allowed them to achieve a take-up rate of 54 per cent.

Over the years, much work has also been done to improve the quality of the food served in South Gloucestershire's schools. Reflecting the Council's stated aim 'to increase the availability and consumption of local food and to promote healthy eating in South Gloucestershire', the Catering Service has undertaken many initiatives to promote local sourcing and to encourage healthy eating practices among schoolchildren (South Gloucestershire Department for Children and Young People, 2005, p7). On the health front, meals served in the schools are mostly home-made and cooked with fresh ingredients (see Figure 6.1). From a nutritional standpoint, menus comply with the nutritional recommendations of the Caroline Walker Trust.[1] Furthermore, as many as 80 schools offer a fruit tuck service with assorted fruit chopped or diced and served in cones (see Figure 6.2).

Primary Lunch Menus & Nutritional Analysis

"Blue" Menu – Week 3

Sample "Blue" Primary Menu

Sept 2005 - February 2006 – Meal Price £1.40

(v) = Suitable for Vegetarians ☺ = locally produced food ⌂ = Home made

	WEEK 3 starting 12th Sept, 10th Oct, 14th Nov, 12th Dec, 16th Jan	
Monday	Oven baked organic burger with gravy ☺ Cauliflower and broccoli bake ⌂ (v) Oven baked organic diced potatoes ☺ Roasted vegetable medley Golden sweetcorn Salad	Chocolate and coconut wholemeal squares ⌂ and peppermint sauce ⌂ Yoghurt / Fresh fruit
Tuesday	Honey roast gammon and gravy ⌂ Cheese and tomato flan ⌂ (v) Creamed organic potatoes ☺ Organic carrots ☺ Garden peas Salad	Rice pudding ⌂ with jam sauce ⌂ Yoghurt / Fresh fruit
Wednesday	Roast beef ☺ with Yorkshire pudding & gravy ⌂ Spicy bean casserole ⌂ (v) Roasted organic potatoes ☺ Broccoli florets Shredded cabbage Salad	Dorset apple cake ⌂ and custard sauce ⌂ Yoghurt / Fresh fruit
Thursday	Chicken pie with gravy ⌂ Vegetarian burger (v) Oven baked sauté organic potatoes ☺ Organic jacket potato ☺ Organic carrots ☺ Creamed swede Salad	Lemon shortbread ⌂ Milkshake Yoghurt / Fresh fruit
Friday	Oven baked cod fish fingers with lemon garnish Tuna pasta bake ⌂ Vegetarian lasagne ⌂ (v) Chipped potatoes Baked beans (low salt & low sugar) Crunchy coleslaw (low fat) ⌂ Salad	Fruit cocktail with ice cream ☺ Yoghurt / Fresh fruit

Figure 6.1 *A sample menu of school meals in South Gloucestershire's primary schools*

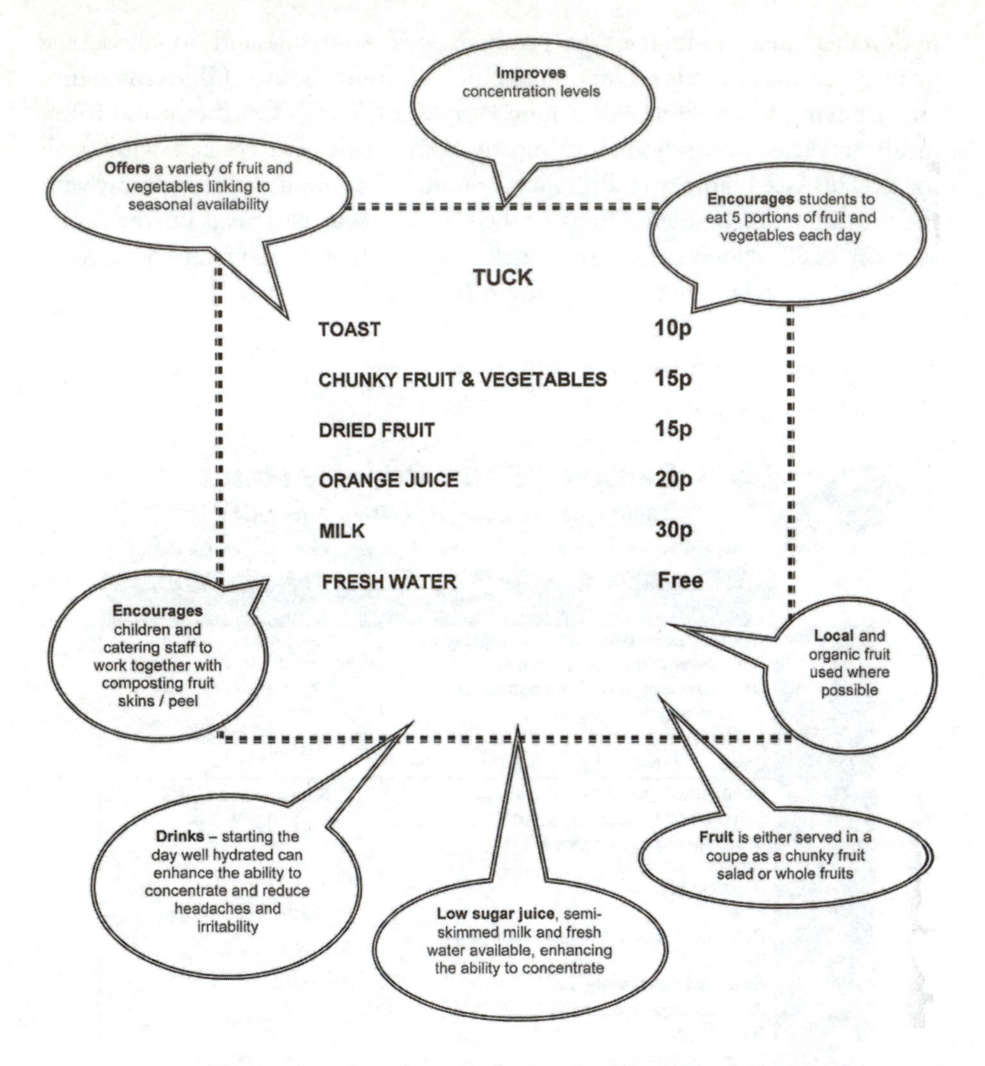

Figure 6.2 *A tuck menu in South Gloucestershire*

The Council has added an explanation of the main benefits associated with the consumption of the various items. Local and organic fruits are used whenever possible.

In addition to encouraging healthy eating[2] and providing a source of energy, the healthy fruit tuck schemes, which are designed in consultation with the schools to link them with their own whole-school approach, have reduced litter on-site. Finally, South Gloucestershire has eliminated or reduced additives, salt, fat and sugar and has banned GM foods from their school menus (South Gloucestershire Department for Children and Young People, 2005).

As for its food sourcing policy, the Catering Service does privilege some local products – for example, beef, pork, turkey, lamb, ice cream and most fruit and vegetables come from within the county. However, in many ways they are moving beyond simple local purchasing. As the Head of Catering explained:

> *We've been doing this for so long, we see a much much bigger picture – much bigger than most people see. [...] We're looking at packaging. We're looking at greenhouse emissions. We're looking at how many vans are on the road. [...] We're looking at how many deliveries we have in a week and whether we've got sufficient storage area so that we can have one delivery instead of three. And do we need all these cans? Do we need all of this?*

In practice, this means that, like in Rome, local foods are complemented with organic, Fair Trade and even conventional products. Tea, coffee, sugar and bananas are Fair Trade. Potatoes and carrots are organic and come from the Prince of Wales's Duchy Farm located in the county. To offset the high costs of fresh organic produce, the Catering Service buys, at a reasonable price, products that would be rejected by supermarkets for merely visual blemishes. This is also a good deal for Duchy Farm. As its manager explained:

> *With potatoes, it's all about skin finish. You get something called 'scabbles' on some varieties, which is just a superficial mark our parents and grandparents would have never even noticed. But because supermarkets want skin that's like an apple skin on a potato, which is very hard to achieve, they were chucking them out. We were getting sometimes 50 per cent chucked out. So we reduced the area that we were growing.*

After making a deal with South Gloucestershire Council to sell them potatoes and carrots at the same price as conventional products, Duchy Farm decided to expand the area cultivated with these two products from 8 to 25 acres.

As for products that are not grown locally or organically, South Gloucestershire relies on 3663 First for Foodservice, a large national food service provider that not many people would immediately associate with sustainable school food. However, the Catering Service has managed to include the company in its school food revolution. As the Head of Catering pointed out:

> *I may not be able to buy local beans, but I can influence what is going on with 3663 in the travelling, the emissions – Are they using green fuel? We were one of the first to say 'we want a combination of provisions and frozen foods', because, you see, we didn't want two vans, one behind the other, turning up. We now have that. [...]*

> *It's not just about what we are trying to do. It's about the influence the service is going to have [...] on people, changing their culture as well as changing your own. It's the ethicalness about saying, 'Why don't you look after your environment? I don't want all this packaging. Do something about it [...]', which again brings us back to the community.*

On the procurement front, this approach has resulted in the development of a quite unique tendering system. South Gloucestershire does not prepare tender documents. Rather, they nominate farmers as 'chosen suppliers' and tender only for the delivery of the food. This New York-style tendering system is good for the farmers, according to the Head of Procurement, because they often do not have the resources to package, weigh, sort out and deliver their products. But most important, it is good in terms of environmental sustainability. As the Head of Catering explained:

> *In the world of climate change and the policy of this Council about emissions and the ethos of the road, it's just far better for us to have one distributor which people deliver to than it is to have 50 different people around the countryside delivering small bags of stuff. [...] It is about things like the emissions, things like [...] how many vans you want in and out of these schools before you end up creating mayhem. So we have done it in a very different way. We have advertised our contracts. We have told them what we want. [...] But underneath this, we bring in the local suppliers through our distributor.*

Generally, farmers involved in the system did not refer to this procurement system as a problem. One of the meat suppliers, for example, identified the problems with supplying South Gloucestershire's schools as the budget constraints and the seasonality of the market, but not the lack of a conventional tendering system. However, there is no doubt that this approach raises significant challenges for some of the local farmers. This is, for instance, the case with a prospective new supplier, who said:

> *We're going to have to stock pile for six to eight weeks [...] and get production up to that capacity. And there isn't a contract agreement – I mean, at the moment it's just a handshake and her word.*

To make this system truly sustainable, South Gloucestershire Council will perhaps have to rethink the informality of its procurement approach and provide its suppliers with a longer-term perspective. As we have seen in the case of Rome, this would allow them to make all the necessary investment that the school food market requires. At the same time, it is important that the spirit of the reform does not get lost. Overall, the story of the school food revolution in this English local authority area is primarily a story of returning dignity to the often forgotten and underpaid workers who make the system function, on a daily basis, through their hard work in the school canteens. In this regard, the main lesson that South Gloucestershire adds to the narrative of the school food revolution is condensed in this anecdote from the Head of Catering:

> *If somebody is having a difficult time in the kitchen, we send out a box of chocolates to them or something. It's not going to be a mega box, because I have to be accountable for public money. But we do try, in our different ways, to say, 'Thank you. Thank you very much.'*

In conclusion, the story of South Gloucestershire highlights that the school food revolution is not just about food producers and consumers. It is also about recognizing that democracy, as a fundamental pillar of sustainable development, will never be achieved without returning dignity and value to those who prepare and deliver the food in the school kitchens and canteens. As the examples of Carmarthenshire and East Ayrshire will show, it is also about creating new forms of social and cultural capital that can *sustain* the process of reform through the difficult times that it often entails.

School Food as Community Development: The Challenge of Local School Meals in Carmarthenshire (Wales)

Good but costly and unproductive: Carmarthenshire's school meal service under the Best Value Regime

Carmarthenshire is a county of roughly 180,000 people located in west Wales. Like many rural areas of the UK, it suffers from problems of poverty and deprivation, which, as is invariably the case, reflect upon the health status of its population. With regard to food consumption habits, it is sufficient to say that more than 58 per cent of this county's residents are overweight or obese – a figure that is 2.3 per cent higher than the Welsh average. With regard to children, recent research found that almost one in ten secondary school pupils do not have breakfast and one in six do not have a cooked meal at home in the evening (Welsh Procurement Initiative, 2005, p18). For many of Carmarthenshire's children, then, school lunches are the main meal of the day.

In this context, Carmarthenshire County Council has made a strong commitment to the school meal service as a tool to improve the eating habits of its population. Indeed when, in 2001, Carmarthenshire's school meal service was subject to a Best Value inspection, it was acknowledged that the service was at a 'good level' on five grounds:

1 Primary school pupils were receiving healthy and nutritionally balanced meals, along with some help to improve their dietary habits.
2 Most secondary school pupils thought that the food served was of good quality.
3 Front-line staff were clear about the objectives of the service and demonstrated customer focus, attention to quality and a common sense of purpose.
4 Paid meal uptake was the highest in the country and free school meal uptake was in the upper quartile for all schools.
5 Much of the service's performance compared well with similar local authorities (Audit Commission, 2001).

In the light of these facts, the Best Value inspectors wrote in their report:

> *We were impressed by the approach of kitchen staff, [...] who were clearly focused on the eating behaviour of the pupils. We came across examples where cooks showed flexibility with the menus to account for the preferences shown by pupils in their care. We heard of situations where pupils who came into the school with less healthy eating habits had been 'converted' to a wider and more healthy range of food.* (Audit Commission, 2001)

Despite these merits, however, the Best Value inspectors also pointed out that Carmarthenshire's catering service was a 'high-cost' service, since pupils in primary schools were paying more for their meals (£1.35 in 2000/2001) than in most other Welsh counties. The 'productivity' of Carmarthenshire's school meal service was also heavily criticized. Employing a metric that seems more attuned to a widget-making factory than a health-promoting school (namely 'meals produced per staff hour'), the Best Value inspectors pointed out that productivity in primary school kitchens was comparatively poor and needed to be tackled (Morgan and Sonnino, 2005).[3] Specifically, they concluded, if productivity could not be improved, and if competitiveness could not be demonstrated, then the Council should engage the private sector in the delivery of the service (Audit Commission, 2001).

We simply need to juxtapose the two pictures provided by the inspectors to realize that they are causally connected: the 'low productivity' in the primary school meals service was a result of the amount of time that

Carmarthenshire's catering staff were devoting to changing the eating habits of their children (Morgan and Sonnino, 2005). The high costs of the service, in turn, related to the Council's commitment to sourcing as much local food as possible for its school meals – an approach that can pay significant economic dividends in poor rural areas, as we will see with the case of East Ayrshire. However, none of these benefits was factored into the desiccated accounting metric of the so-called 'best value' audit. As one representative from the Council succinctly put it:

> *The auditors were looking at numbers; what we were interested in was the quality and the nutritional value. They were not competent enough to value that.*

Challenging the conventional development metric: Carmarthenshire's reform of the school meal service

By asking the County Council to become a lower-cost producer, the Best Value process could have easily led 'to the perverse spectacle of higher productivity alongside a lower health dividend' (Morgan and Sonnino, 2005, p30). Fortunately, however, Carmarthenshire managed not just to defend its school meals system from the criticism of the inspectors, but also to start a process of improvement of the service. As a member of the Council explained, referring to the aftermath of the inspection:

> *All of a sudden there was a very clear understanding of what the authority is about. From the procurement angle, for the first time we had that clarity of objective [...], we could see that [...] we could do something more than just buy goods and services.*

Crucial to the success of the reform was the Council's ability to enlist cross-party political support around a broader metric of sustainable development that informs the vision of the county's Community and Corporate Strategy – one of the most innovative and 'joined-up' local government strategies in the UK (Carmarthenshire County Council, 2004). Through its 2004–2020 Community and Corporate Strategy, Carmarthenshire has committed itself to 'tackling the causes of ill health by looking at life in the round'. Another key theme of the strategy is 'regeneration', which was taken to mean, among other things, promoting sustainability within the county by supporting local producers and suppliers (Cullen, 2007, p223).

In an effort to work 'thematically, rather than departmentally', as the Head of Catering explained, a multi-agency partnership, which included Carmarthenshire County Council, the County Council Lifelong Learning and

Leisure Department, the NHS Trust, the Local Public Health Team and Carmarthenshire Healthy Schools Scheme, was formed, and in April 2004 a School Meals Strategy was launched. In its objectives, the strategy integrates the different goals of sustainable development. Health and wellbeing are promoted through weekly menus that comply with the proportions contained within the 'balance of good health' – 33 per cent fruit and vegetables, 33 per cent cereals and potatoes, 15 per cent milk and dairy products, 12 per cent meat and fish, and 7 per cent foods containing sugar and fat. In consultation with parents and children, new menus were produced that provided traditionally cooked, fresh and seasonal food, with reduced additives, salt and sugar and no GM-ingredients (see Box 6.1). Significantly, Carmarthenshire also decided to remove choice in primary schools – an initiative that, as the Head of Catering pointed out, contributed to improving pupils' behaviour after lunch, as reported by many head teachers.

Box 6.1 *An example of Carmarthenshire's weekly menu in primary schools*

Monday
Home-made cheese and tomato pizza
Baked beans
Thick cut chipped potatoes
Fruit yoghurt or fruit bowls

Tuesday
Home-made shepherd's pie
Peas and carrots
Boiled potatoes with gravy
Peaches and custard

Wednesday
Roast Welsh lamb with mint sauce
Broccoli and carrots
Boiled and roast potatoes with gravy
Fruit and ice cream

Thursday
Home-made chicken Korma
Mixed white and brown rice
Peas
Naan bread
Fruit and chocolate sponge pudding and custard

Friday
Roast gammon with pineapple
Green beans
Jacket or boiled potatoes
Yoghurt jelly

Source: Carmarthenshire County Council (2005)

By adopting a 'whole-school approach', the strategy also emphasizes the need to raise children's awareness of the importance of food to good health and to promote the educational values of the dining experience, which should also target, the document points out, children who bring packed lunches to school. In this regard, it is worth noticing that Carmarthenshire places a special emphasis on issues of social inclusion. Indeed, catering staff are explicitly told to 'treat all pupils equitably and fairly and ensure that no one is discriminated against' (Carmarthenshire Partnership, 2004, p7). To this end, the document also specifies that caterers must provide for the needs of children on therapeutic and vegetarian diets.

From an economic perspective, the School Meals Strategy stresses the potential of school food to develop a more sustainable economy. The document in fact encourages caterers to support, wherever possible, local producers and the rural economy (without, however, forgetting companies that practise other 'ethical procurement policies' such as Fair Trade). In this respect, the School Meals Strategy has been linked to, and supported by, another strategy: the Local Sustainable Food Strategy, launched in November 2004. Designed 'to build sustainable development into Carmarthenshire County catering food contracts', this strategy provides an inspiring and comprehensive model for sustainable procurement that challenges the misconceptions and shortcomings associated with public food provisioning discussed in Chapter 2.

Building on the principles of sustainable development (which the Welsh Assembly Government has a constitutional obligation to promote[4]), the Local Sustainable Food Strategy sets out the following key principles and objectives:

- adoption of *'Food for Thought'* as the guiding principle for food procurement in Carmarthenshire; this principle 'maximizes the development of local supply chains where appropriate and encourages whole-life costing to identify best value for money' (Carmarthenshire Catering Services, 2004, p6);
- provision of *nutritious meals* and promotion of health and wellbeing through the use of quality ingredients in line with Carmarthenshire's Nutrition Strategy;
- *looking after the environment* and natural resources whilst minimizing waste;
- achieving *continuous improvement in the catering service* across all levels;
- adoption of *ethical procurement policies* that include organic and Fair Trade; and
- development of *food specifications* that take into consideration the nutritional content, freshness, shelf life, traceability and packaging of the products as well as their ethical qualities (for example organic, Fair Trade and PGI status) and their contributions to sustainable development.

To involve small suppliers in the procurement process, the Council organized a briefing day for existing and potential suppliers to explain its aims, expectations and policies. Significantly, in compliance with the principles of sustainability, the Local Sustainable Food Strategy anticipates that these aims and policies:

> *may include a requirement to conserve energy, promote biodiversity, use agricultural practices that are less reliant on fossil fuels and agri-chemicals, support animal welfare, minimize packaging, use energy-efficient appliances, recycle oil, plastics, cans and glass, and generally consider long-term environmental impacts.* (Carmarthenshire Catering Services, 2004, p10)

From rhetoric to reality: The challenge of local sourcing in rural Wales

Translating these principles into practice has not been easy in a local authority that provides 19,000 school meals each day across a total of 145 schools. As a member of the Council explained:

> *The very small businesses were all saying the same to us: you are a very difficult animal to deal with, you are very big and we don't know who to talk to.*

The Council has been making special efforts to involve small local producers. For example, they have attempted to facilitate the inclusion by large contract holders of small growers and producers as second- and third-tier suppliers; they have promoted the use of lots during the tendering process to allow small and medium-sized enterprises to bid for parts of the supply contracts; and they have trained producers through formal and informal meetings and initiatives (Cullen, 2007, p224). However, small food producers in Carmarthenshire seem to be having major difficulties in meeting the demands of the school food market. In commenting on her attempt to promote cooperation amongst three cheese producers, who would need to join forces to provide the 23 tons of cheese the schools consume every year, the Head of Catering said:

> *There's a huge market here, but the businesses have not yet developed; they are not yet sophisticated enough to meet our demand.*

Presently, all fresh meat, bottled water, fruit juices, bacon, eggs and sliced meat are sourced within Wales. All of the ice cream, about half of the bread and a significant proportion of the milk are sourced from within the county, and as far as possible fresh vegetables are also obtained locally (Cullen, 2007, p224). Children are enjoying this school food revolution. Indeed, as the Head of

Catering explained, during the first year of the reform, the consumption of fruit and vegetables has increased by 40 per cent and that of milk has doubled.

However, there are threats to the economic sustainability of Carmarthenshire's school food revolution. Indeed, the Council had to face a significant increase in expenditure for sourcing higher-quality ingredients – an increase that reached 75 per cent for fresh meat. At the same time, the overall budget for school meals has also shrunk since the Best Value inspection by as much as £400,000.[5]

In short, the story of Carmarthenshire's school food revolution is not yet a story of local food procurement. Rather, it is a story of procuring locally – in other words, relying upon local suppliers, but not necessarily upon local products. Clearly, much work still needs to be done to improve the economic sustainability of the school food market and to turn it into a practical tool for local community development. In this sense, East Ayrshire's pilot initiative to revolutionize the food served in the county's primary schools has important lessons to teach to the rest of the UK and the many rural areas of Europe and North America that are attempting to utilize school meals as a tool to re-localize their struggling economies.

The Power of the Local: Sustainable School Meals in East Ayrshire (Scotland)

Reinventing school meals as a health service: The context

The county of East Ayrshire, in west central Scotland, shares much in common with Carmarthenshire. It is a deprived rural area of 120,000 residents, with an unemployment rate of 4.5 per cent and a large number of households (more than 6 per cent of the total) dependent on benefits. As in Carmarthenshire, this deprivation affects the health status of East Ayrshire's population: 40 per cent of all households in the county have one or more people with a long-term illness (Sonnino, 2007c).

In contrast with Carmarthenshire, however, East Ayrshire's aspiration for school food reform could count on a level of political and financial support from the national government that was unprecedented in the UK.[6] In the late 1990s, devolution gave Scotland primary and secondary legislative powers over many areas related to food policy. Social justice became a priority of the newly formed Scottish Government, which immediately began to take initiatives to improve health and reduce health inequalities.[7] In this context, expenditures for the improvement of public health in Scotland went from £2 million per year in 1996–1997 to £86 million per year in 2005–2006 (Lang et al, 2006, pp10–12).

For the first time in the UK, school meals were reinvented as a health and wellbeing service, rather than a commercial one. Indeed, as we discussed in Chapter 5, *Hungry for Success* introduced a whole-school approach to the provision of school food, using social inclusion principles as important policy drivers (Lang et al, 2006, p49). To make all Scottish schools 'health promoting' schools by 2007, the report adopted new nutritional standards for school food, encouraged links between learning and teaching about healthy eating in the curriculum and food provision in the schools, and promoted the active involvement of pupils and staff in the reform process (Scottish Executive, 2002). Significantly, as much as £63.5 million was committed to support school food reform between 2003 and 2006 and a similar level of funding (£70 million) was allocated for a further three years in 2006. Against these strengths, *Hungry for Success* 'had one glaring weakness: it made little or no reference to how and where the food was produced' (Sonnino and Morgan, 2007, p130). One of the factors that makes East Ayrshire so distinctive is the fact that this local authority has overcome the inherent weaknesses of *Hungry for Success* by incorporating a mix of organic, Fair Trade and locally produced foods into its very innovative school food reform.

In addition to reforming its school food policy, Scotland is also committed to creating a regulatory context favourable to a sustainable approach to public procurement. Bearing in mind the barriers to sustainable procurement we identified in Chapter 2 and the limitations of the UK's national procurement policy we described in Chapter 5, two factors are especially worth mentioning here. First, in 2004, the Scottish Executive released its *Sustainable Procurement Guidance for Public Purchasers*, which states that buyers can legitimately specify requirements for freshness, delivery, frequency, specific varieties and production standards (Scottish Executive, 2004). Second, in 2007, the Scottish Parliament hosted the first ever debate on green procurement in the UK, which revealed striking cross-party support for environmentally friendly procurement practices (Sonnino and Morgan, 2007, p131). This stimulating debate provides an ideal opportunity to contextualize the real and potential implications of East Ayrshire's school food reform. In fact, pointing to the fears and uncertainties that surround public procurement, the Green Party's Eleanor Scott said:

> *At present, a procurement officer who wants to improve his or her green buying practices needs extraordinary personal dedication and initiative. [...] Everybody looks at the East Ayrshire food example with awe and wonder as well as with admiration, but that approach should be the norm.* (Scottish Parliament, 2007)

Seizing the opportunity: East Ayrshire's school food reform

In the spirit of 'joined-up' thinking, and with the new funding provided by *Hungry for Success*,[8] in 2004 East Ayrshire's Head of Catering decided to see 'whether it was possible to create a sustainable school meal system' (Gourlay, 2007, p212) by trialling the Soil Association's Food for Life standards (see Box 6.2). In the UK, these are widely regarded as the gold standard of school food reform because of their holistic approach to healthy and locally sourced products.

Box 6.2 *The Food for Life recommendations for school meals*

- School lunches must meet the nutritional standards identified by the Caroline Walker Trust.
- At least 75 per cent of the food consumed over a week must be made from unprocessed ingredients.
- At least 50 per cent of the ingredients used for each meal must be locally sourced.
- At least 30 per cent of the food should be from certified organic sources.
- Classroom education on food, cooking and farming must be promoted.

Source: Soil Association (2003)

In August 2004, the Head of Catering introduced a pilot scheme at Hurlford primary school, based on the use of fresh, organic and locally sourced foods, which gained the support of children, parents, and catering and teaching staff – so much so that the uptake of school meals immediately increased. On the basis of these results, in early 2005 the Head of Catering made a case to the Council's Education Committee for a wider roll-out of the initiative. As in Carmarthenshire, East Ayrshire's Council sees school meals as a 'cross-cutting' service that is closely linked with three main themes of the community strategy: health (specifically, the need to address health inequalities through lifestyle changes); economic sustainability (particularly the need to address rural depopulation through the development of the local economy); and environmental sustainability (the promotion of local food products). As the Chief Executive Officer explained:

> *[The school meal service is about] improving the environment [...] and improving opportunities. [...] We have a huge problem with population loss [...], so anything you can do in terms of contributing to the sustainability of the economy has a positive impact on our community, and in terms of local produce, that's [also] a contribution to the environment. So we see [school meals] as being very cross-cutting.*

In 2005, the Council decided to extend the pilot to another 10 primary schools in the county. Under the stated objective of ensuring that the ingredients used in the schools were 'nutritionally valuable, fun and interesting' and introduced children to a wider range of foods than many would find at home (East Ayrshire Council, 2005, p28), the Head of Catering formed a specialized team, composed of a dietician, a head teacher and the school catering managers, and hired a Scottish food stylist to assist them in devising radically new recipes. In the new four-week menus:

- the amounts of potatoes, pasta, rice, fruit and vegetables were increased to promote the health benefits of the Mediterranean diet;
- fat, sugar and salt were reduced and replaced with natural flavours, herbs and spices;
- added colourings, artificial flavourings and GM-foods were banned; and
- fresh and unprocessed ingredients were prioritized.

As in Rome and New York City, East Ayrshire's reform was based on a very inclusive procurement approach that actively involved all actors in the food chain. On the supply side, before drafting the tender documents the Council held a series of open meetings with local food producers to informally explain the aims of the initiative and provide guidance on tendering (Anthony Collins Solicitors, 2006, pp77–78). As in Carmarthenshire, it became immediately clear, as one procurement officer explained, that:

> *One of [the small suppliers'] major concerns was that they couldn't beat big farmers on price. So we showed them how we actually evaluated; we told them that they actually had to concentrate on other areas as well.*

On the basis of these findings, the Council devised a very creative tendering process that aimed to involve organic and small suppliers in the school meals system. In particular:

- To attract organic suppliers, some of the strict 'straightness' guidelines for Class 1 vegetables were made more flexible. Within the specification, it was also stated that 'as the purpose of the pilot is to introduce fresh/organic food into school meals, the organic alternative should be offered as far as possible' (East Ayrshire Procurement Section, 2005, Section 2.1).
- To enable small local suppliers to compete with larger companies, the tendering contract was divided into nine lots (instead of the four previously used): red meat; dry, bottled and canned foodstuffs; fruit; vegetables; milk; cheese; eggs; fish; and poultry.

- To improve the quality of the ingredients served in the county's schools, award criteria were equally based on price and quality. To reward suppliers' various contributions to sustainable development, quality was broken down into four sub-criteria (see Box 6.3).

Box 6.3 *Creative procurement in practice: East Ayrshire's quality criteria for the awarding of the school meal contract*

'Quality', which comprises 50 per cent of the award criteria in East Ayrshire, refers to:

- *ability to supply to deadlines (15 per cent)* – suppliers' adaptability in terms of delivery methods and their proposed timescale from harvest to delivery;
- *quality and range of foodstuffs (15 per cent)* – suppliers' capacity to provide Fair Trade, seasonal and traditional products; the shelf life of their products; traceability and recall procedures; their quality system as approved by an accredited certification body; product assurance schemes; inward inspection procedures; procedures for inspection during manufacturing; equipment inspections; suppliers' capacity to meet ethnic, cultural and religious diet needs;
- *food handling arrangements and facilities (10 per cent)* – safety in the working environment, training opportunities for staff, equality issues, food safety management, membership of food associations and so on; and
- *use of resources (10 per cent)* – suppliers' contribution to biodiversity, initiatives adopted to minimize packaging and waste (for example reuse, recycling, composting), and compliance with animal welfare standards.

Source: East Ayrshire Procurement Section (2005)

The Council notified the tender details to all suppliers who had been involved in the first pilot scheme and adverts were placed in the local press and in the official journal of the EU, anticipating an overall value of the food to be purchased of about £120,000. The Council received a total of 13 bids and, in October 2006, it awarded contracts to seven local suppliers – two wholesalers and five producers.

As in Carmarthenshire, East Ayrshire's producers had to face challenges with the school food market. The organic fruit and vegetables supplier, for example, had to add an extra delivery line a week and a whole extra packing shift a week to meet the demand of the 12 schools – which, however, represented just 5–10 per cent of his market revenues. The small organic dairy that supplies 800 litres of milk to East Ayrshire schools every week is struggling with the seasonality requirements of the school food market. The local artisanal cheese-maker had to invent a new 'basic plain cheese' that is mild enough to suit children's taste but also hard enough to keep for weeks, thereby allowing her to cut the number (and costs) of deliveries.

Despite these difficulties, in East Ayrshire suppliers are very committed to the school food market. As the meat supplier pointed out:

> *The business we do with East Ayrshire is very important to us in terms of dealing with a local customer; we don't want to lose that customer [...]. We need to look out for each other. Children are the future of any company. Children are the future.*

Similarly, the artisanal cheese-maker supplying East Ayrshire's schools said:

> *I like the school [market], because [...] if you educate [children] to good eating, then it starts to affect the whole structure of the economy. Later on, when they grow up and have children, it gets passed on.*

East Ayrshire Council has worked hard to also create the same kind of commitment amongst other actors in the food chain. Training sessions on nutrition and healthy eating have been organized for catering managers and cooks. Suppliers have been invited to enter the classroom to explain to children where and how they produce food. Parents have also been taken on board through a series of 'healthy cooking tips demonstrations'.

Through the adoption of this inclusive and creative procurement approach, East Ayrshire has managed to exceed the Food for Life targets. In 2006, 50 per cent of the ingredients used in the county's schools (fruit, vegetables, milk, flour, pulses, pasta, couscous and brown rice) were organic, about 70 per cent were locally sourced (including bread, cheese, red meat, poultry and eggs) and more than 90 per cent of all food on the menu was unprocessed (Gourlay, 2007, p216). The reform has also been producing important outcomes from a sustainable development perspective. Environmentally, as a result of the Council's local-sourcing approach, the average distance travelled per menu item has decreased from 330 miles (in the standard menu) to 99 miles – cutting food miles by 70 per cent (Gourlay, 2007, p216).[9] Anecdotal evidence from catering staff also suggests that the new approach has produced less packaging waste in the schools. Economically, the reform has created new opportunities for local suppliers, while at the same time delivering a multiplier effect of £160,000 for the local economy. Socially, the reform has increased customer satisfaction with the service: 67 per cent of children think that school meals now taste better, 88 per cent of children agree that they like fresh food and 77 per cent of parents believe that the scheme is a good use of the Council's money. Moreover, school staff feel that the 'localness' of produce is improving the quality of the service and is creating positive links amongst schools, communities and the environment (Bowden et al, 2006, p2). There was also a small but encouraging rise in uptake of school meals at the onset of the reform, but almost no increase in labour costs arising from the local foods initiatives.

As for the costs of the ingredients, local suppliers offered prices that were on average 75 per cent higher than those offered by larger national firms, but the effects within the schools came down to just 10 pence per meal. Parents faced no increase in the price paid (£1.52/meal in 2007): in fact, larger portion sizes and higher nutritional quality have been funded through subsidies provided by the Council and by the *Hungry for Success* programme.

In the light of these positive results, in May 2007 the reform was extended to 26 schools in East Ayrshire. By June 2008, the Council had included in the reform more than 30 primary and secondary schools. However, this initiative entails all kinds of new challenges for the local authority. At the production end, feedback provided by local suppliers suggests that work still needs to be done to overcome prevailing perceptions that bids are evaluated on the basis of cost, rather than quality, and that the public food market requires too much time and bureaucratic work (Bowden et al, 2006, p62). Moreover, as the project expands, small suppliers face the need to step up their systems to meet requirements related to greater production costs, more deliveries, more production and more administration (Bowden et al, 2006, p42). To afford these changes, as we have seen in the case of Rome, producers will need to secure longer contracts than the single-year ones awarded in 2004.

As mentioned above, *Hungry for Success* will remain in place until 2009, a time span that probably will not be sufficient to guarantee the kind of cultural change necessary to sustain school food reform. Indeed, at the consumption end of the food chain, after the initial rise in take-up rates, East Ayrshire has recently experienced an average reduction of 1.5 per cent in the number of school meals served in its primary schools. It is still too early to understand whether this negative trend is just a transitional challenge or whether we are dealing with a backlash against the promotion of a healthy food culture (Sonnino and Morgan, 2007). However, in a country where the 'freedom of choice' ideal provides even the youngest consumer with an opportunity to purchase cheap and unhealthy food at local fast-food shops, the sustainability of East Ayrshire's school meals system will depend mostly on the Council's capacity to win over the hearts and minds of parents to the cause of healthy eating. In other words, the future of school food reform in this poor area of Scotland is mostly linked to the authority's ability to help local residents understand, as the Head of Catering stated, that with 'much of this [...] redirected into the local economy, these changes deliver a net benefit to the community' (Gourlay, 2007, p216).

The Rural Revolution in School Food Provision: Lessons and Challenges

The three rural areas we have discussed in this chapter capture different, but equally important, dimensions of the school food revolution. Yet in many ways, the narratives form a continuum around 'civic' and 'domestic' quality conventions that emphasize place-making and community-building. South Gloucestershire is a story of re-qualification of the core of the school meal service: labour. In this relatively wealthy English local authority, the main driver of the reform was the desire to return dignity, recognition and respect to the underpaid workers who run the service on a daily basis. In South Gloucestershire, public authorities have understood that these workers are not just dinner ladies: they are resource managers and health workers in disguise. Building on the ideals of democracy and social justice in the workplace, this rural area has also managed to promote economic development and environmental integration – in other words, sustainable development.

Social values also lie at the heart of Carmarthenshire's school food revolution. Confronted with the neo-liberal logic of the Best Value regulatory regime, this deprived rural area of Wales adopted a 'joined-up' approach that explicitly emphasizes the multiple benefits of school food for the community as a whole – in terms of the population's health, rural regeneration and local economic development. Much still needs to be done in Carmarthenshire to fully embed the school meal service in the local environmental and socioeconomic context. However, there are already two important lessons that can be drawn from this case study. First, the story of Carmarthenshire highlights the need for more public debate about the *metric* that informs accounting conventions and audit trails in the monitoring of sustainable school food systems. Indeed, the Welsh case study clearly shows that, far from being neutral devices, the values and assumptions embodied in that metric can foster or frustrate re-localization strategies. Second, the school food reform initiated in Carmarthenshire also underlines the need for political support to embed healthy school meals into a wider strategy for sustainability and community wellbeing. In simple terms, then, school food reform cannot take place in a vacuum. To be effective and sustainable, local innovations need to be supported by complementary action at the upper echelons of the multi-level polity, as in Scotland, where another deprived rural county, East Ayrshire, has managed to design one of the most creative school food systems in the world.

Building on the financial and political support provided by the national government, East Ayrshire has reconnected producers and consumers in the name of environmental sustainability. In East Ayrshire, producers, policy-makers, catering managers and consumers are working together to create and sustain 'the local'. Considering the tangible results that East Ayrshire has so far achieved, the importance of this shared commitment should not be underestimated in the quest for sustainable development. This applies not only to the many rural areas of Europe and North America that are attempting to devise development strategies to temper the negative effects of decades of agricultural modernization – it also applies to the poorest areas of the developing world, where, as we will discuss in the next chapter, a new era of school feeding is now emerging: 'home-grown'.

7

Home-Grown: The School Feeding Revolution in Developing Countries

The formula is simple: food attracts hungry children to school. And education broadens their options, helping to lift them out of poverty. When we launched the campaign, we knew that serving nutritious food at school would improve attendance and school performance. What we didn't project was just how important School Feeding would become as a platform for launching major health initiatives. 'Each child is everyone's child.' – this East African saying captures the thinking behind the Global School Feeding Campaign. (WFP, 2004)

If obesity is the main concern of school food reformers in the developed countries of Europe and North America, chronic hunger is the chief concern in the developing countries of Africa, Asia and Latin America. Of all the dramatic contrasts between rich and poor nations today, none is more compelling, or indeed more tragic, than the fact that the number of obese people in the world is now roughly similar to the number of hungry people. At the beginning of the new millennium, some 1.2 billion people were estimated to fall in each category, underlining the fact that malnutrition comes in many shapes and sizes.

It is difficult to describe the numbers of hungry and undernourished people as anything other than shocking and shameful. According to the United Nations (2007), we have the following situation:

- Over 1.3 billion people live on less than US$1.00 a day and some 854 million people in the world are chronically undernourished.
- More than 6 million children die every year from hunger.
- Some 351 million school-age children are chronically hungry.
- More than 115 million children are not enrolled in school, the majority of whom are girls.[1]
- Roughly one-third of all children in developing countries are stunted or underweight.

Overall, some 96 per cent of school-age children who are not enrolled in school are located in developing countries, and almost three-quarters of these children live in sub-Saharan Africa and South and West Asia, areas where the most intractable problems of poverty, hunger and underdevelopment are concentrated.

The fact that chronic hunger affects so many men, women and children in the 21st century makes a mockery of a 'right' that was first proclaimed 60 years ago. It was in 1948, in the Universal Declaration of Human Rights, that the UN first formally recognized the right to food as a basic human right. The right to food is the right of every person to have regular access to sufficient, nutritionally adequate and culturally acceptable food for an active, healthy life. It is the right to feed oneself in dignity, rather than the right to be fed. The right to food, then, is a legal obligation as well as a moral imperative (FAO, 2007a).

This basic human right has been reaffirmed time and again by the international community, most notably at successive World Food Summits. Considering the record of governments to date, however, it must be said that the most basic of all human rights is the most consistently and universally violated (George, 1990). At the World Food Summit in 1976, for example, 150 governments solemnly pledged that, within a decade, no child would ever go to bed hungry – a pledge they signally failed to honour.[2] Twenty years later, at the World Food Summit in 1996, more than 180 governments committed to eradicating hunger and, as a first step, they set an intermediate target: to halve by 2015 the number of undernourished people in the world from the 1990 level. On current trends, however, this looks like another empty pledge. In fact, more than ten years after the event, the Food and Agriculture Organization of the United Nations (FAO) was forced to acknowledge that 'we are confronted with the sad reality that virtually no progress has been made towards that objective' (FAO, 2006, p4). In its authoritative report *The State of Food Insecurity in the World*, the FAO (2006, p4) went on to say that:

> *This publication has highlighted the discrepancy between what could (and should) be done and what is actually being done for the millions of people suffering from hunger. We have emphasized first and foremost that reducing hunger is no longer a question of the means in the hands of the global community. The world is richer today than it was ten years ago. There is more food available and still more could be produced without excessive upward pressure on prices. The knowledge and resources to reduce hunger are there. What is lacking is sufficient political will to mobilize those resources to the benefit of the hungry.*

As we will attempt to demonstrate in this chapter, school feeding programmes (SFPs) have a crucial role to play in the fight against poverty and hunger. If properly designed and supported, financially as well as politically, they provide a unique strategic platform on which campaigns can be fought to turn rhetoric into reality. This is especially evident in the context of the recent shift from conventional school feeding to *home-grown* school feeding, a shift that signals nothing less than a quiet revolution.

School Feeding and Food Aid: The Role of the World Food Programme

SFPs have existed in one form or another for more than 50 years; they are funded and managed by a wide range of organizations, including governments, NGOs and aid agencies. Although the aims depend on the context, conventional school feeding has been defined in generic terms as 'a set of interventions supporting both medium-term nutritional and long-term education objectives that are being implemented with food as the primary resource' (Bennett, 2003, p7). Since the World Food Programme (WFP) is by far the largest provider of SFPs around the globe, we will focus on the scope and limits of its role and initiatives.

The WFP school feeding system

As the food aid arm of the UN, the WFP was scheduled to start operating as a three-year experimental programme in January 1963. In reality, however, it started its relief work at the end of 1962, because of the crises triggered by an earthquake in Iran, a hurricane in Indonesia and the return of 5 million refugees to Algeria after the war.

The fact that the WFP has no funds of its own renders it totally dependent on voluntary donations from governments, corporations and individuals, with some 60 governments providing the bulk of the funding. Donations come in three main forms – cash, food (flour, beans, oil, salt and sugar, for example), and items to grow, store and transport food. These resources have to be spread across the WFP's three main types of activity:

1. emergency programmes to cover natural and man-made disasters;
2. protracted relief and recovery operations to get disaster-hit areas back on their feet; and
3. development programmes to stimulate longer-term social and economic development by providing workers with rations to build vital infrastructure and by offering children food aid as a reward for going to school.

The WFP is required to devote at least 50 per cent of its development assistance to least developed countries and at least 90 per cent of it to low-income, food-deficit countries. To help it to target specific areas and schools, the WFP uses the following criteria:

- identify food-insecure areas;
- select the food-insecure areas with the most urgent educational need (for example, low enrolment/attendance, high gender disparity and high drop-out rate);
- assess the feasibility of forming effective partnerships with national and international agencies working in the area;
- evaluate the capacity to guarantee a minimum level of hygiene and safety (especially for female staff and students);
- understand the prospects for strong community involvement, through, for example, the creation of a parents and teachers association;
- assess the ability to store and prepare food adequately; and
- evaluate the likely cost-effectiveness of the project.

In the WFP school feeding system, two forms of food provisioning are used: in-school feeding and take-home rations. As regards in-school provisioning, each menu is chosen according to local tastes and customs, nutritional needs, local foods available, ease of preparation and resources available. There are usually four meal options: breakfast, mid-morning snack, lunch and, for boarding schools only, dinner. The timing and nature of the meal depends on the length of the school day, the local customs and the availability of trained cooks, a well-equipped kitchen and clean water, among other things. A sample of menus for each type of meal is given in Table 7.1.

The second form of provisioning, the take-home ration, is designed to accommodate the fact that, in very poor countries, children are often required to help their families make a living (for example by working in the fields, guarding livestock, caring for younger siblings, gathering firewood or searching for food). This means that some children do not have the time, the economic means or the energy to attend school. When on-site school feeding is impossible to implement or insufficient to reach particularly vulnerable children, such as girls or orphans, the WFP provides take-home rations of basic food items – a sack of cereals or several litres of vegetable oil, for example – which are distributed to families to help offset the loss of the child's contribution to the family's livelihood.

Between 2002 and 2005, the WFP studied more than 1 million children in over 4000 schools across 32 countries in sub-Saharan Africa to assess the impact of its school feeding activity. It found that its SFPs had a strong impact on absolute enrolment in WFP-assisted schools, especially during the first year of programmes, when the average enrolment increased by 28 per cent for girls and by 22 per cent for boys. After the first year, however, increases in absolute

Table 7.1 *Sample menu for in-school feeding*

Type of Meal	*Sample Menu*	*Total Kilocalories*	*Country Example*
Breakfast	Corn-soya blend 8g Porridge 8g Sugar 8g Vegetable oil* 8g	404	Timor
Mid-morning snack	High energy biscuits 100g	450	Iraq
Breakfast and lunch	Maize meal 150g Pulses 40g Vegetable oil* 10g Salt 3g	955	Rwanda
Breakfast, lunch and dinner	Cereals (rice or maize) 450g Pulses 45g Canned fish 25g Vegetable oil* 20g	2027	Benin

* Vegetable oil fortified with vitamin A.
Source: WFP (2006)

enrolment were found to vary substantially by the type of SFP (WFP, 2006, p13).

Since food alone is not enough, the WFP has developed what it calls the *Essential Package*, which aims to deliver complementary nutrition, health and basic education in conjunction with partner agencies like UNESCO, UNICEF and the World Health Organization. During 2005, for example, the Essential Package interventions focused on micronutrient supplementation and systematic de-worming.

The WFP's school feeding initiatives attach a high priority to the involvement of parents through School Committees.[3] Daily cooking of food is usually performed by mothers or other community members, who cook for the children – very often on an unpaid basis – throughout the school year. Although parental and community involvement is a vital ingredient in the WFP school feeding recipe, this is not always easy to achieve, as we will see in the case of Ghana.

To deliver its SFPs, the WFP seeks to build effective partnerships with a whole series of other organizations – governments, other UN agencies, NGOs, the private sector and, of course, the local communities in which it works.[4] When it is assured that the domestic government has the capacity and the resources to manage the SFP itself, the WFP seeks to develop an exit strategy.[5] The main criteria that inform a WFP exit strategy concern the following:

- *Milestones for achievement:* The time and conditions for the exit need to be identified at the programme design stage and must be communicated to (and agreed with) all the stakeholders.
- *Inclusion of the private sector:* The WFP believes that active private-sector involvement helps to develop support and expertise among key political and economic players.
- *Management and communication:* The WFP ensures that the programme leadership is taken over by national actors and the exit plan is understood, especially by teachers, parents and beneficiaries.
- *Government commitment:* Exit is more successful if it involves contributions from the host government – in other words, an active role for government in implementation and capacity development.
- *Community involvement:* A commitment from the community, especially the parents, is essential; it is important that they contribute from the beginning, with cash or in-kind support.
- *Technical support:* Technical support is needed throughout the project, during the phase-out process and beyond; this is particularly important for ensuring an adequate transfer of skills and maintaining the benefits long after external assistance has ended (WFP, 2006, p22).

The reach of WFP school feeding has increased dramatically in recent years – from 11.9 million children in 52 countries in 1999 to 21.7 million children in 74 countries in 2005. The WFP's current plan is to reach 50 million children by the end of 2008 (WFP, 2006).

The scope and limits of conventional school feeding

The SFPs of the WFP are arguably one of the great humanitarian success stories of our time. For all its achievements, and despite its very laudable aims, however, conventional school feeding is not above criticism. The most important criticisms revolve around three issues:

1. the nature of the health impact;
2. the extent of educational benefits; and
3. the fact that food aid, when it is donated in-kind, can undermine poor farmers by substituting imported food for local food.

The long-term health impact of conventional school feeding is not very clear for two main reasons. First, the vast majority of SFPs are, by their very nature, partial in coverage, so they do not include children who are not in school, arguably the most vulnerable child populations of all. And second, if, as a good deal of evidence suggests, the first two years of a child's life are the most important in terms of its development, school feeding is addressing the wrong age group. Where poverty is endemic, the physical growth of children aged six to nine is affected mostly by environmental factors, including poor diet, illness, lack of sanitation and poor hygiene. The potential for catch-up growth among stunted schoolchildren is thought to be limited after two years of age. Furthermore, stunting at two years has been shown to be associated with later deficits in cognitive ability, a fact that highlights the need to act early (Mendez and Adair, 1999; Bennett, 2003; World Bank, 2006).

If the health benefits of school feeding are less clear-cut than they appear, for some critics the same can be said of the supposed educational benefits. The long-term food security objectives of school feeding are based on two presumed linkages:

1. that school feeding improves educational outcomes; and
2. that the resulting increase in literacy and numeracy will have a positive impact on food security through increased productivity, higher employment, better natural resource management, higher incomes, smaller families and so forth.

Although there is a lot of evidence to support the latter, the former linkage appears to be more problematical. School feeding does indeed provide a crucial incentive to get children to enrol, attend and continue in school; however, the mere presence of the student does not guarantee positive educational outcomes. In other words, enrolment, attendance and retention are necessary but insufficient components of improved education. To overcome this criticism, SFPs must meet some minimum level of education standards. In short, there must be 'an emphasis on quality education and education *outcomes* rather than inputs' (Hicks, 1996).

The third criticism concerns the negative effect of food aid on poor farmers in developing countries. This is perhaps the biggest problem of all, so an in-depth discussion of this problem is crucial to fully understand the potential of school feeding as a development tool.

The paradox of food aid: A barrier to development?

The effectiveness of aid is a hotly debated subject within the international donor community. Critics allege that the vast bulk of foreign aid has been a failure, not least because it was concerned more with satisfying the political priorities of Western donors than with the pressing needs of poor countries. To illustrate this great food aid paradox – where aid actually harms the very countries it is designed to help – we focus on the US, the country with the longest history of food aid and the largest food aid programme, and also the most honest about the dual role of food aid as a geopolitical weapon as well as a humanitarian instrument. No one understood this dual role better than Herbert Hoover, who was a master practitioner of this food aid philosophy long before he became the 31st President of the US. As one critic has written:

> *Herbert Hoover was the first modern politician to look upon food as a frequently more effective means of getting one's own way than gunboat diplomacy or military intervention, and as a means of supporting US farmers into the bargain. He first came on the scene at the beginning of the First World War. Hoover sold US wheat and summarily forced settlements of several European disputes simply by threatening to cut off food aid to the party of whose politics he disapproved.* (George, 1986, p193)

American food aid policy was eventually formalized in Public Law 480, also known as the Food for Peace Law, which was passed by the US Congress in 1954 and aimed 'to increase the consumption of US agricultural commodities in foreign countries'.[6]

Criticisms of American food aid policy have grown louder and louder in recent years, not least because the US is the only country to utilize what donors call 'monetized food aid', a method by which grain is shipped from the US to charities in the developing world (to American charities like CARE, Save the Children and World Vision, in particular), who then sell the grain in local markets to finance their own feeding programmes. Having been the leading distributor of American food aid for the past 50 years, CARE finally broke ranks in 2007, when it said it would no longer use monetized food aid. David Kauck, a Senior Advisor at CARE, explained this highly significant move, which meant that the charity was turning its back on US$46 million a year in US federal funding, by saying:

> *If we are trying to limit people's vulnerability to food insecurity, we just couldn't see how we could continue [monetized food aid] in good faith.* (Harrell, 2007)

No less damaging to the credibility of American food aid policy was a Government Accountability Office (GAO) report published in 2007, which exposed its inefficiencies and lack of effectiveness. In fact, although it recognized that the US is the world's largest food aid donor, accounting for over half of global food aid at an annual cost of US$2 billion, the GAO discovered that burgeoning business and transportation costs had triggered a 52 per cent decline in the average tonnage of food delivered between 2001 and 2006. Even more disturbing, these business and transport costs were absorbing as much as

65 per cent of the total food aid budget, which meant that only one-third of the money allocated to food aid was actually spent on the food itself.

Although the Bush administration has sought to reform the food aid system, for example by suggesting that a quarter of the budget be used to buy food in developing countries in emergencies, this modest proposal has been defeated in Congress by a powerful coalition of agri-business, shipping and charity interests, each of which has a deep financial stake in the current system. As things stand, the American food aid industry is protected by a law that requires 75 per cent of all its donated food to be grown in the US and shipped on US flag carriers employing American crews. As well as being extremely expensive, this means that the system is also very time-consuming, with the whole process taking up to six months. Not surprisingly, the GAO concluded its report by saying that the US food aid system was not getting the right food to the right people at the right time (GAO, 2007)[7] (see Figure 7.1).

At the receiving end of the food aid process, in the hunger hot spots of developing countries, the criticisms of US policy are even more trenchant. In Malawi, one of the poorest countries in the world, Oxfam's Mary Khozombah sees the food aid paradox at first hand. As she stated:

> *People who want to help Malawi need to support agriculture by educating farmers and improving irrigation and help people find other forms of income. We need empowerment so our farmers can export. Ask us! We might come up with good ideas. Food aid should be the last resort, in an emergency – and even then it should be bought locally if possible. Of course, if people say we want to give you food, we'll say yes – you can't say no. Poor nations like us too often just accept the charity without looking properly at the effects. But in the long term it really kills our people.* (Renton, 2007)

According to one of the most influential critics of the international food aid programme, hunger and poverty in developing countries will be solved not by foreign aid, but rather by 'home-grown' reforms from within (Easterly, 2006). Proponents, on the other hand, maintain that aid, when it is intelligently designed and delivered, has made a genuine difference to the quality of life in the beneficiary countries (Sachs, 2005). To get beyond these polarized claims and counter-claims it is worth focusing on three key issues in the debate – the *level* of aid, the *nature* of aid and the *governance* of aid.

No one has done more to expose the woefully inadequate *level of aid* than Jeffrey Sachs, the author of *The End of Poverty*. One of the great merits of the book is that it effectively demolishes the myths about foreign aid that continue to resonate in many developed countries today, particularly in the US. For example, the former US Secretary of the Treasury, Paul O'Neill, expressed a

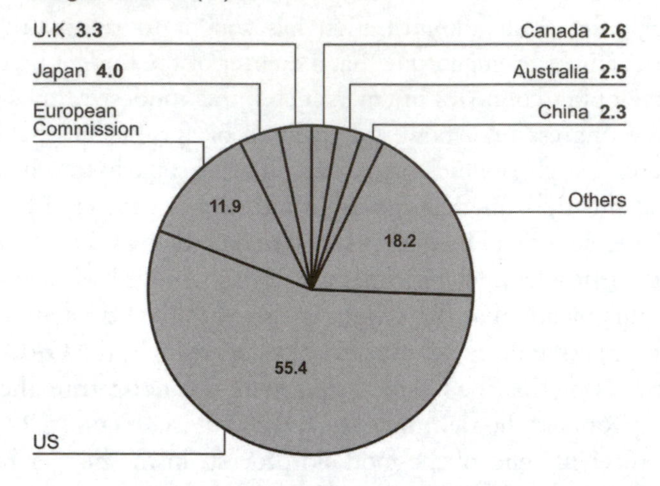

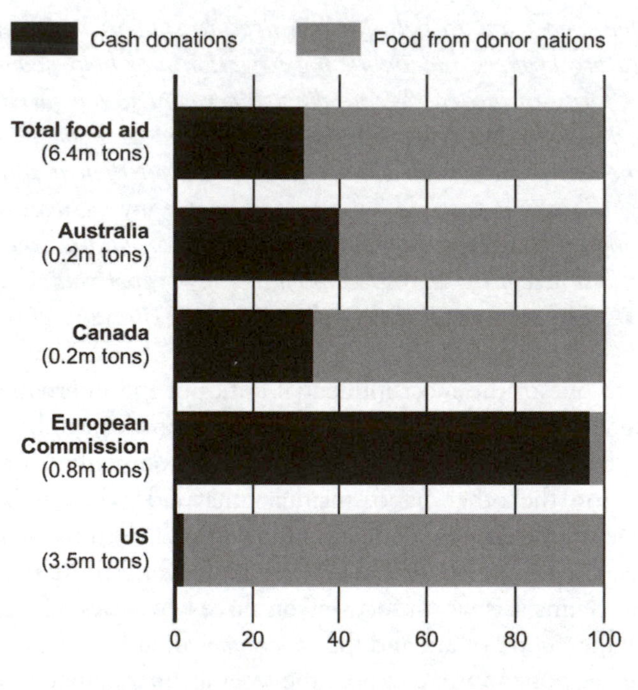

Figure 7.1 *Food aid policy*

Source: Beattie (2008)

common American frustration when he said, in reference to aid for Africa, 'We've spent trillions of dollars on these problems and we have damn near nothing to show for it.' Sachs argues that the main reason why there is little to show for the aid to Africa is 'because there has in fact been so little aid to Africa!' (Sachs, 2005, p310). Contrary to popular belief, the amount of aid per African per year is extremely small – just US$30 per sub-Saharan African in 2002 from the entire world. Indeed the actual sum is even smaller, because, of that modest amount, nearly US$5 was allocated to consultants from donor countries, more than US$3 was for food aid and other emergency aid, another US$4 went towards servicing Africa's debts and US$5 was for debt relief operations, leaving just US$12 that actually went to Africa. As for the US share of foreign aid, which was US$3 per sub-Saharan African in 2002, Sachs shows that just *six cents* actually reached the person after all the deductions were made. The conclusion, he says, is clear: 'if we want to see the impact of aid, we had better offer enough to produce results' (Sachs, 2005, p310).

Rich country governments committed themselves in 1970 to a UN target to provide 0.7 per cent of gross national income as aid, a pledge that the vast majority of them failed to honour. Indeed, the only countries to have honoured the pledge are Sweden, Luxembourg, Norway, The Netherlands and Denmark. With the exception of Greece, the US is positioned at the bottom of the league table of overseas development assistance as a percentage of gross national income, a problem compounded by the contrast between its military and aid budgets. In 2004, for example, the US was spending *30 times* more on the military than on foreign aid: US$450 billion to US$15 billion (Sachs, 2005, p329). Unlike the G8 group of rich countries, the EU has now agreed a timetable to reach the UN target of 0.7 per cent of gross national income by 2015. Meeting international commitments on the level of aid – without fiddling the figures by including such dubious items as debt relief, military training and loosely defined technical assistance – is essential if foreign aid is to deliver more tangible results.

While critics accept that the level of aid is less than what has been pledged, some of their most forceful criticisms concern the *nature of aid*, particularly the paternalistic and neo-colonial relationships between donors and beneficiaries. Foreign aid, be it money or food, has been a geopolitical weapon since time immemorial, as we saw earlier, and this tendency certainly did not end with the Cold War: for example, two-thirds of all US aid goes to Israel and Egypt, its key allies in the oil-rich Middle East. In this context, another problem, according to a 30-year veteran of the World Bank, is 'donor paternalism', which helps to explain the failure of many aid projects in Africa (Pomerantz, 2004).

Donor paternalism can be counterproductive in many ways: it can undermine the success of a project; it can embitter a bilateral relationship; and, by preventing donors from recognizing and respecting local knowledge, it can blind them to the home-grown solutions that Africans, for example, have devised for themselves. The challenge here is for developed countries to embrace developing countries as genuine partners who have something to offer, rather than as supplicants in a donor–recipient relationship that is doomed to remain unequal. The quintessential requirement for a more balanced partnership is for developed countries to recognize, and respect, the fact that African countries are agents of their future, rather than totally powerless victims of circumstance (Morgan et al, 2007a).

The third issue concerns the *governance of aid*. From the donor side, the current governance system is profoundly inefficient, since it imposes very high transaction costs on the beneficiary. One of the most important issues here is the question of aid harmonization. In the past, poor countries were obliged to negotiate with scores of different donors, each of whom may have had a separate accounting system, creating an enormous burden on a public administration system that could be using its limited capacity to better effect than hosting visiting delegations and preparing quarterly progress reports for each donor. What developing countries need above all else is direct budget support to scale-up public investments, and for this money to flow from a single multilateral donor, like the World Bank, rather than from a multiplicity of bilateral donors (Sachs, 2005).

Beneficiary governments, for their part, also need to do more to promote good governance. To get more direct budget support, developing country governments will need to demonstrate to the donor community that they are giving sufficient priority to poverty reduction and that they operate open and transparent public finance systems. Without such assurances, donors have the right to be sceptical of the promises of developing country elites, some of whom have been known to plunder the public purse for personal gain or divert national resources from constructive to destructive ends. As Sanchez et al (2005, p8) note:

> *Highly food-insecure countries tend to spend two to three times as much on defence as on agriculture [...]. This combination of declining investment in agriculture and rising military expenditures is extremely worrying – and a sorry indicator of the real priorities of governments and donor agencies, despite their stated commitments.*

Developing countries invariably record low scores in the league table of good governance. The annual Corruption Perceptions Index operated by Transparency International (see Chapter 2) gives the impression that developing countries are more corrupt than developed countries, but Transparency International concedes that future rankings will have to provide a more balanced message because 'countries that have less corruption internally very often continue to play a major role to perpetuate corruption in poorer parts of the world' (Williamson, 2007).

Thus reforms in these three areas – in the level, nature and governance of aid – would help to make the aid process more effective, more efficient and more equitable. However, what all these changes overlook is the fact that aid, while it dominates the development agenda, is really only a small part of the picture. To suggest that foreign aid has a modest role to play in fostering development in poor countries is not at all to suggest that aid is ineffective, as the anti-aid lobby argues. Rather, it is to recognize that the developed world pursues a wide range of policies that have an impact on the developing world. Foreign aid policies are arguably the most visible, but they may not be the most important. Paul Collier has produced one of the most compelling arguments along these lines. As he points out, aid can be effective only if it is embedded in a genuine development strategy that harnesses a wide array of other policy instruments:

> *Our support for change can be decisive. But we will need not just a more intelligent approach to aid, but complementary actions using instruments that have not conventionally been part of the development armoury: trade policies, security strategies, changes in our laws and new international charters.* (Collier, 2007, p192)

Collier is especially concerned with what he calls the 'bottom billion', that is, the very poorest countries of the developing world. As he argues, aid can do little or nothing to help these countries if they remain afflicted by internecine conflict, if they cannot implement standards of good governance or if their elites are allowed to appropriate public funds for private purposes with the tacit support of developed country banks. Equally important, according to Collier, the trade policies of the rich donor countries – particularly the EU and the US – can only be described as *anti-development*, because they make it difficult for poor countries to access Western markets, while their subsidized exports hurt poor farmers in developing countries. In short, Collier is calling for more synergy between the different policies (e.g., trade, aid and security) of the developed countries.

Clearly, the failure to reduce extreme hunger can no longer be explained by lack of resources, lack of knowledge or lack of ability. Today, the only credible explanation for the existence of hunger in the midst of plenty is the lack of *political will*, particularly, though not exclusively, in the developed countries. However, a new constellation of forces is emerging, composed of new global trends as well as new home-grown philosophies in developing countries, and this looks like the kiss of death for traditional food aid, which has paradoxically damaged the long-term food security prospects of the very countries it was ostensibly designed to help.

From Aid to Development: The Home-Grown Revolution

Signalling a whole new era in the history of school feeding and development aid, the home-grown revolution has its origins in a unique combination of global and local circumstances. At the global level, the most important factor was the Millennium Summit in 2000, where the UN member states collectively committed themselves to the eight Millennium Development Goals (MDGs) shown in Box 7.1.

Box 7.1 *The Millennium Development Goals*

1. Eradicate extreme poverty and hunger.
2. Achieve universal primary education.
3. Promote gender equality and empower women.
4. Reduce child mortality.
5. Improve maternal health.
6. Combat HIV/AIDS, malaria and other diseases.
7. Ensure environmental sustainability.
8. Develop a global partnership for development.

The MDGs immediately raised the status of school feeding. In fact, SFPs directly address the goals of reducing hunger by half, achieving universal primary education and promoting gender parity. They also contribute to the reduction of disease and poverty, providing a platform for promoting child health, environmental education and the prevention of major diseases like HIV/AIDS and malaria.

A new approach to reducing hunger inevitably involved a new commitment to agriculture, given that over 70 per cent of all undernourished people live in rural areas and depend directly on agriculture for their food and livelihood (United Nations, 2007). Given the enormous significance of the agricultural sector to the poorest of the poor, it beggars belief that foreign aid to agriculture has been allowed to decline so steeply in recent years. Clearly, agriculture had fallen victim to the fads and fashions that infect donor policies from time to time. In fact, it was deemed to be an irredeemably 'backward' sector of the economy, unable to compete with the more fashionable nostrums that urbanization and industrialization are the 'motors' of modernization.

The bias against agriculture in the international aid community was eventually exposed for what it was: an unwarranted ideological fashion. 'Many studies have shown', the FAO stated, 'how agricultural growth reduces poverty and hunger, even more than urban or industrial growth' (FAO, 2006). What best captured the new, pro-agriculture turn in development thinking was the

UN-sponsored report *Halving Hunger: It Can Be Done*, the product of the UN Task Force on Hunger (Sanchez et al, 2005). In addition to its seven recommendations, *Halving Hunger* contained two highly pertinent criticisms of conventional aid policy:

1 that poverty reduction strategies in the past had paid too little attention to agriculture and nutrition; and
2 that when they were used, agricultural and nutritional interventions tended to be implemented separately, diminishing their overall impact.

To overcome the poor design of poverty reduction strategies, the task force called for a more integrated approach that addressed the *causes* of hunger and not merely the symptoms. This was the context in which the task force recommended that 'all feeding programmes be sourced, where possible, from locally produced foods rather than imported food aid' (Sanchez et al, 2005, p12). To be effective, the task force said, interventions needed to be deployed 'synergistically'; that is to say, when two or more such interventions are combined, the overall effect is greater than the sum of their individual effects. As an example of synergistic action, the task force identified three major initiatives to combat hunger:

1 community nutrition programmes;
2 home-grown SFPs; and
3 soil health and water programmes.

As Sanchez et al (2005, p18) state:

> *A combination of the three interventions may constitute an attractive, new, integrated programme in rural areas facing the dual challenge of high chronic malnutrition and low agricultural productivity. Community nutrition and home-grown school feeding programmes can be initiated in tandem with basic investments in soil and water. The increased local production will have a ready market in the home-grown feeding programmes. The resulting synergies of better education outcomes (particularly for girls), greater agricultural production and incomes, and improved nutrition for mothers and babies will address a community's hunger from multiple angles – opening the way for other interventions.*

The *Halving Hunger* report firmly established the need for a twin-track approach to hunger reduction – an approach that alleviates the symptoms in the short term and helps to address the causes by raising agricultural productivity and promoting rural development. With the role of agriculture having been rehabilitated in the repertoire of poverty reduction strategies, the mainstream international development bodies began to absorb the new climate of opinion. Even the World Bank, a bastion of conservative thinking on development, had an epiphany and became pro-agriculture.[8]

The twin-track approach to poverty reduction has worked in many developing countries, according to the FAO, the biggest exception being sub-Saharan Africa, 'the region with the highest prevalence of undernourishment, with one in three people deprived of access to sufficient food' (FAO, 2006, p5). Looking ahead to 2015, the FAO predicts that the states of this region will be home to some 30 per cent of the undernourished people in the developing world, compared to 20 per cent in 1990. Sub-Saharan Africa is also the main locus of 'the bottom billion', the poorest of poor people, whose development prospects have been stymied by a noxious combination of domestic factors (like civil war and political corruption) and global factors (like unfavourable terms of trade for basic commodities) (Collier, 2007). In this context, home-grown measures hold an especially promising potential, and it is for this reason that the rest of the chapter will focus on sub-Saharan Africa, where, as we will see, food insecurity is more a political than a purely agricultural problem; and where, therefore, the twin-track approach has to be complemented by a range of other domestic measures.

In addition to the global factors that created a new climate of opinion, local factors were also at work within the African developing countries. Perhaps the most important new factor in the African context was the New Partnership for Africa's Development (NEPAD), which was adopted at the 37th summit of the Organization of African Unity (OAU) in 2001 with the following objectives:

- to eradicate poverty;
- to place African countries, individually and collectively, on a path of sustainable growth and development;
- to halt the marginalization of Africa in the globalization process; and
- to accelerate the empowerment of women.

Significantly, the first of NEPAD's founding principles was a commitment to *good governance* as the basic requirement for peace, security and sustainable development. Indeed, there is no doubt that poor governance can exact a terrible price, turning food-secure states into food-insecure ones almost overnight.[9]

Clearly, however, NEPAD is more an aspiration for the future than a reflection of current political conditions in Africa. Moreover, NEPAD is essentially an elite-driven initiative, and the international donor community has been too eager to forge partnerships with the leaders rather than with the peoples of Africa. But despite these shortcomings, NEPAD signalled the first time that Africa's elite collectively committed itself to the principles of good governance.

It also committed the continent's leaders to new partnerships – internally, with business and civil society groups, and externally, with the international aid community. Against this background, NEPAD signed a memorandum of understanding with the WFP in 2003, and home-grown school feeding (HGSF) was identified as one of the priorities for early action, with Ethiopia, Ghana, Kenya, Malawi, Mali, Mozambique, Nigeria, Senegal, Uganda and Zambia agreeing to be the first wave of countries to pilot the new programme (WFP, 2007a). Indeed, school feeding was even considered to be the single most important long-term investment towards the reduction of poverty and food insecurity in Africa.

Home-grown school feeding: The key challenges

According to Emmanuel Ohene Afoakwa, the secretary of the African Network of School Feeding Programmes, the biggest single challenge facing the home-grown model is how to sustain it following the exit of the WFP. Governments in developing countries find it difficult to assume full responsibility for the management of the home-grown programme because of the shortfalls of governance, finance and procurement (Morgan et al, 2007a). Let us briefly examine each of these problems in turn.

Governance

Broadly speaking, governance has vertical and horizontal dimensions: the vertical dimension refers to a multi-level governance system that embraces supra-national, national and sub-national levels of political organization; the horizontal dimension refers to inter-organizational relationships at each of these levels. If it is to be successfully implemented, HGSF requires good vertical and horizontal relationships: vertically, between central and local government, so that the latter can deliver what the former has decreed; and horizontally, between government and its partners in business and civil society. Far from being a wholly government-controlled process, HGSF, as we have seen in other chapters, needs to mobilize many different stakeholders: if the latter are to collaborate for mutually beneficial ends, they require a reliable institutional framework in which they can have some confidence.

Finance

If governance is a perennial challenge for developing countries, access to stable sources of finance is even more problematic. Two of the biggest financial challenges are the reform of the international food aid system and the stability of national budgets in developing countries. The WFP would ideally like its donations to be in the form of up-front untied aid. If the US can be persuaded to switch from food aid to cash aid, the WFP would have more opportunities to finance the home-grown model, at least during its all important teething stage.

Within the developing countries the key challenge is to enable national governments to assume the financial responsibility for managing the HGSF model. As there are so many competing claims on the national budget of a poor country, the international donor community could play a very important role here by contributing directly to these budgets on condition that the domestic government ensured that the money did not 'leak' from constructive purposes (like nutrition projects) to destructive purposes (like military projects).

Procurement

The twin challenges of governance and finance beset all SFPs. But the home-grown model faces a third challenge in the form of procurement, because, unlike conventional school feeding, HGSF aims to use the power of purchase to substitute locally produced food for food imports. Procurement managers in developed countries have the luxury of knowing that their suppliers are skilled enough to meet exacting purchasing specifications. Their counterparts in developing countries, by contrast, face a far more difficult procurement task: instead of browbeating suppliers to get the lowest possible price or the best value, they have to fashion a supply side from scratch, relying upon small-scale farmers who are working in the unremittingly hostile and precarious world of subsistence agriculture.

Procurement has never been consciously deployed to promote local development in developing countries. Indeed, of all the humanitarian aid spent in Africa in recent years, a mere 10 per cent has been spent on locally produced goods and services (Simpson, 2006). As the largest procurer of food aid in the world, the WFP is acutely conscious of its duties as a socially responsible purchaser, so much so that it is making strenuous efforts to convert its own food procurement budget into a development tool. To understand the potential of its procurement power, let us briefly look at the financial scale and geographical scope of the WFP's food purchasing activity.

A rough measure of the *scale* of WFP food procurement activity can be gauged from the fact that in 2006 it procured 2 million metric tons of food in 84 countries, worth over US$600 million. The geographical *scope* of this activity was such that (in tonnage terms) 77 per cent was purchased in developing countries and 50 per cent in the least developed and low-income countries. In 2006, the WFP made most of its food purchases in Africa: as Table 7.2 shows, four African countries feature in the top ten food origin countries.

The bald statistics help us to understand the scale and scope of WFP purchasing, but they tell us nothing about the principles that inform the decisions as to what is bought, where and why. These procurement principles are summarized in Box 7.2.

Table 7.2 *Top ten food origin countries, 2006*

Rank	Country	Food purchased (million US$)
1	Uganda	41.2
2	Ethiopia	37.0
3	Pakistan	34.7
4	Canada	32.5
5	Kenya	29.7
6	South Africa	28.6
7	Ecuador	28.0
8	Turkey	27.9
9	Indonesia	22.5
10	Belgium	22.0

Source: WFP in Africa

Box 7.2 *The WFP food procurement process*

The main objective of the WFP's food procurement, as derived from its Financial Rules, is 'to ensure that appropriate food commodities are available to the beneficiaries in a timely and cost-effective manner. Consistent with this, WFP purchases must also be fair and transparent.' In addition, it is stated that 'when conditions are equal, preference will be given to purchasing from developing countries'.

The WFP principally engages in three different levels of food procurement: local, regional and international. International procurement is largely effected by the Food Procurement Service at WFP headquarters through tenders that request varying quantities of commodities for multiple destinations. Consequently, the tenders often cover a large geographic scope, which can include both developed and developing countries.

Regional procurement is carried out through tenders for offers for a specific region. These tenders are principally launched by WFP Regional Offices. Local procurement is mainly effected by WFP Country Offices and refers to purchasing within a recipient country for beneficiaries in the same country.

Purchases are made from pre-qualified suppliers through the competitive bidding process. Vendor selection criteria include:

- the legal capacity to enter into a contract;
- a specialization in the particular commodity;
- the financial standing necessary to honour a contract; and
- proof of previous satisfactory performance.

Tender invitations call for specific quality standards, delivery terms, packaging and markings.

The above information raises the following question: How does the WFP decide where to buy? A donation may be accompanied by a donor's request to procure in a certain manner or area, or for a particular destination, which obligates the WPF to procure the commodities as requested.

If a contribution does not come with a donor condition attached, then the decision on where to purchase is driven by the primary objective of food procurement: to provide appropriate food to the beneficiaries in a timely and cost-effective manner. In many instances, purchasing locally has the considerable advantage of reducing the cost of transport as well as the time needed to get the commodities to the beneficiaries. In addition, taste acceptability is higher. Consequently, in the majority of cases local food procurement harmoniously achieves its primary objective and effects a resource transfer to the economies of recipient countries. However, to ensure that every purchase fits our main objective, a cost comparison is made, on a case-by-case basis, between the locally available option and what it would cost to bring in the same commodity from the regional or international markets.

The results of the tenders themselves are often a very good reflection of the various markets within which we operate. However, they cannot always provide a full picture of every market. An understanding of the texture of agriculture and food security in any given country serves as the most important base from which a market profile is built. The WFP's procurement network undertakes constant market analysis as it recognizes that a knowledge of the main crops, agro-ecological zones, levels of production, agricultural seasons, latest food balance sheets, size, location and importance of food markets, principal exports and imports, major barriers to the free flow of food, and inter-regional trading and transport patterns is mandatory.

Food procurement is a very important link in the supply chain that brings assistance to those most in need as rapidly and efficiently as possible. The above shows just how essential this function is and the efforts that are made to procure food quickly and responsibly.

Source: WFP (2008)

Honed for more than 40 years, the procurement expertise of the WFP contrasts markedly with the procurement capacity of the countries in which it works, creating a knowledge gap that has to be bridged if the agency wants to make a smooth exit. Creating local procurement capacity in a developing country context is an extremely difficult task, however. As we will see in the next section, the main challenge is that supply and demand must be nurtured in conjunction with one another.

The African Pioneer: The Ghana School Feeding Programme

Of the African countries that committed themselves to the home-grown model of school feeding, the West African state of Ghana has made all the early running. Known as the Gold Coast in colonial times, Ghana won its freedom from the British in 1957 and celebrated 50 years of independence in 2007. Despite occasional bouts of political instability in the past, Ghana is now considered to be one of the most stable and best-governed states in Africa, not least because it has had four consecutive free and fair elections. Seasoned political observers of the African scene believe that the critical election occurred in 2000, when a *peaceful* transfer of power was effected from the National Democratic Congress (NDC) to the New Patriotic Party (Kwame, 2007).

With a population of 22 million, Ghana has made real progress in reducing the number of people who live below the national poverty line; indeed, in national terms poverty has fallen from 52 per cent of the population in 1991/1992 to 28 per cent in 2005/2006. However, this encouraging national picture conceals some major regional disparities, with the most acute levels of poverty and food insecurity in the three northern regions – Northern, Upper East and Upper West – where nearly 70 per cent of the poor live (WFP, 2005).

In terms of its governance system, Ghana is divided into 10 regions and 138 local districts, with the capital city, Accra, located in the extreme south of the country. As a result of a recent programme of decentralization, these regional and district levels of administration are now expected to assume more responsibility for the welfare of their areas. This process carries both threats and opportunities. On the positive side, decentralization might foster local knowledge and local accountability at the district level, enabling local communities to play a more active role in the management of their affairs, which is crucially important to the success of the home-grown model. On the negative side, however, the richer and better organized districts will be best placed to make the home-grown model work for them, perhaps further deepening the territorial divide between the poor north and the richer south of the country.

SFPs have a long history in Ghana thanks to the pioneering work of two organizations in particular: Catholic Relief Services (CRS) and the WFP. The SFP of CRS has been operational in Ghana since 1959 and uses imported food aid from the US. It targets deprived schools in northern Ghana and the cost works out at roughly US$0.15 per student per day. One of the best features of this programme is that it involves a high degree of community participation: School Management Committees have responsibility for all matters related to the school and Community Food Management Committees are responsible for overall management – a division of labour that leaves the teachers free to focus on teaching. CRS also provides training for the community to improve management, sanitation and cooking methods. Once the community is trained to manage the programme, the role of CRS is merely to transport commodities to the school. For its part, the WFP has been managing SFPs in Ghana for some 40 years and it ensures that its targeting of schools is coordinated with CRS and the Government of Ghana. Equally important, it ensures that its targeting criteria are transparent and are validated by the communities in which it works.

Originally, the new HGSF in Ghana was to be funded as a joint venture between the Ghanaian Government and NEPAD, but the latter was unable to deliver its share of the agreement. Fortunately, the HGSF programme was able to go ahead because donor support from the Dutch Government enabled the Government of Ghana to cover the early costs of the programme. Formally launched as a four-year programme in 2006, the Ghana School Feeding Programme (GSFP) has three major objectives:

1 to reduce hunger and malnutrition;
2 to increase school enrolment, attendance and retention; and
3 to boost domestic food production. (GSFP, 2006)

The total budget for the four-year programme is estimated to be around US$211.7 million, nearly 90 per cent of which is food costs. It is also expected that other collaborating institutions – national ministries and district assemblies, for example – will commit resources in cash or in kind to complement the official budget. 'The GSFP programme,' said a Dutch NGO, 'when properly funded and implemented as designed, has the potential to change the hunger, education, and ultimately the food security and poverty landscape in Ghana for good' (Sign, 2006). These were bold claims for a programme that faced the three challenges of finance, governance and procurement identified earlier in this chapter.

To manage the programme, a new set of governance structures were put in place at national, regional, district and community levels, but these new arrangements never worked as intended. At the national level, the government formed a ministerial oversight committee composed of five different ministries covering health, agriculture, education, women and children, and local government. Since all these ministries had a legitimate interest in the fate of the GSFP, the prestigious new programme triggered a furious power struggle at the very top of the national government. Currently, the Ministry of Local Government, Rural Development and Environment has formal responsibility for the programme, and there is logic to this, given that the GSFP needs to be implemented at the level of the district.

At the district level, the District Chief Executive, a government appointee, is supposed to form a district implementation committee (DIC) in each of Ghana's 138 districts. Officially, the DIC is expected to take 'absolute control' of the programme in its district, including the opening of bank accounts and the procurement of foodstuffs (GSFP, 2006). In many cases, however, the DICs do not function and no actual meetings are held, with the result that the District Chief Executive forms a 'one-person committee', as the WFP puts it, which takes decisions on behalf of the whole community.

At the most decentralized level – the school – a school implementation committee (SIC) is supposed to recruit the cooks for the beneficiary school, determine the menu, procure the food from local farmers and oversee the cooking and the feeding. Although the SIC is designed to have core members, such as parents and local chiefs, some head teachers have to do all the work themselves to ensure that the children are actually fed.

The original governance arrangements for the GSFP were laudable and sound, designed as they were to enable decision-making to be decentralized and transparent, as shown in Figure 7.2. In reality, however, the programme has fallen far short of the original design. As the WFP team was forced to conclude after its field visit:

> *The programme has been implemented in a way that meant that key stakeholders were not adequately prepared to play their respective roles at the beginning of the programme. Before programme implementation, the GSFP was supposed to launch sensitization and education programmes for these stakeholders to explain to them the objectives of the programme, implementation guidelines and their roles. This did not happen, and as a result, the stakeholders implementing the programme have little or no understanding of what they are implementing or their respective roles. [...] Decentralization must create opportunities for the governed to take decisions that affect their wellbeing. This is largely not happening in the implementation of the GSFP. As a result, decision-making is sometimes personalized and mostly centralized in the hands of a few. In the event, decision-making has taken a top–down approach.* (WFP, 2007b, p44)

If some of these problems stemmed from a failure to deliver what had been designed, other problems arose from poor design. Of all the design faults, perhaps the most important, and most debilitating, was the original decision to create *new* governance structures instead of using the old structures that were already in place. Despite the existence of district assemblies and school management committees, which could (and should) have substituted for the newly minted institutions at district and school levels, 'the GSFP ignored all these legally constituted bodies and established new bodies with no legal mandate' (WFP, 2007b, p45).

Procurement has proved no less of a problem than governance. In the original design of the programme, it was thought that the SIC, composed of local people, would collaborate with local farmers to ensure that local supply also meant local foodstuffs. In the event, two other procurement systems have emerged, sometimes called the 'supplier model' and the 'caterer model', both of which could compromise the goal of creating local markets for local farmers.

Outsourcing of food procurement to an outside supplier was a new development that began in 2006, when the GSFP was scaled up after its more

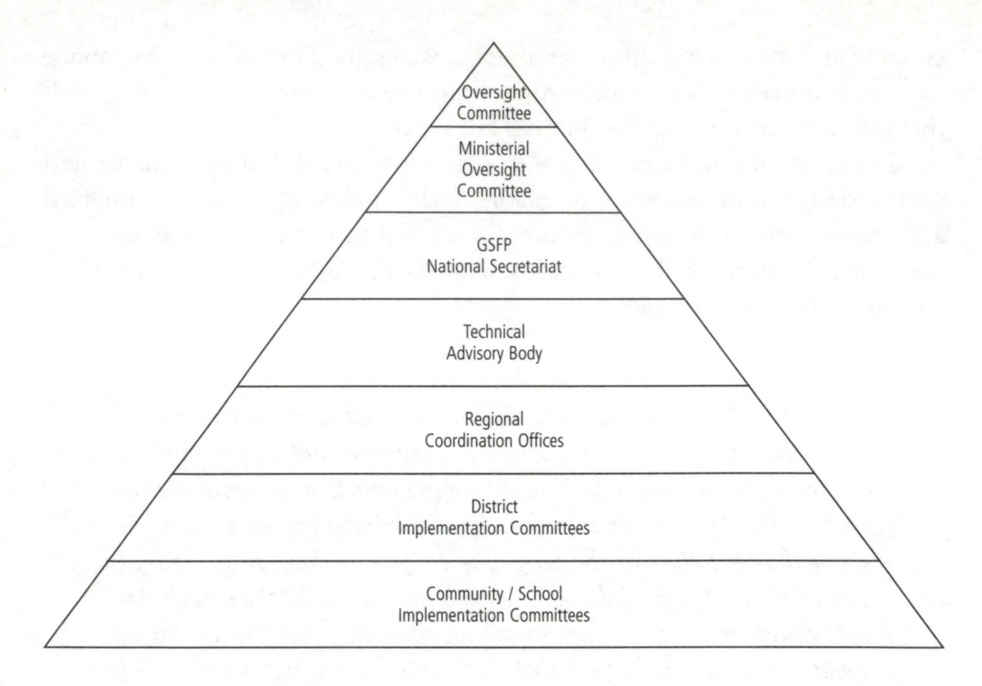

Figure 7.2 *Multi-level governance of the Ghana School Feeding Programme*

Source: WFP (2007b)

modest pilot phase. The attractions of the supplier model of procurement are twofold: teachers can focus on teaching and, because of their greater access to cash and credit, private suppliers can often reduce the delays associated with the public spending process. However, as the WFP case study shows, the supplier model has some serious shortcomings. By buying outside the beneficiary communities, the private supplier provides no direct market opportunities for local farmers, thereby defeating one of the central objectives of the GSFP. The supplier model also devalues the role of the local community, which has no involvement in deciding what food is procured and from where.

The caterer model has been implemented in urban areas, like the Greater Accra region, where communities are more apathetic about being involved in school-based activities. In this model, the caterers fulfil all the key functions: they procure the food, store it, cook it in a central kitchen away from the school site, deliver it to the school and finally serve it to the children. Although the menus are agreed with the district assembly, there is little or no involvement of the school or the local community. This caterer model has the same shortcomings as the supplier model, with the additional problem that urban caterers appear to buy more imported products, undermining not just local farmers in the vicinity of the school, but domestic farmers in general.

Such are the shortcomings of these models that the WFP team decided to recommend what they called *a school-based procurement model*, in which 'the community decides what to buy, when to buy and the cost' (WFP, 2006, p28). Furthermore, since there is no commercial middleman or long-distance movement of goods, the school-based model appears to be more cost-effective and transparent. Despite all the advantages of the school-based model, however, it seems that some district officials are inclined to frown upon it because its informal character could 'open the door to corruption' (Fisher, 2007, p22).

Although the practice of the GSFP has fallen short of the promise, some encouraging results did emerge from the pilot phase, especially as regards enrolment, which increased by 20.3 per cent in beneficiary schools (against 2.8 per cent in non-beneficiary schools). Retention was also better in beneficiary schools. Success, however, breeds problems of its own, because larger classes are not matched by extra teachers, with the result that the quality of education could be undermined, unless new resources are forthcoming to manage the growth in pupil numbers.

The WFP has shown that the elaborate governance arrangements that were originally designed to manage the programme are clearly not working. In particular, the ministries of agriculture and health are not making much of a contribution, more needs to be done to improve monitoring and evaluation, and, above all, 'the communities must be given the opportunity to contribute to the programme' (WFP, 2007b, p53).

The decentralized school-based procurement system that was part of the original design of the programme has been bypassed, according to the WFP field study. The use of the supplier and caterer models of procurement helps to explain why the GSFP has failed to make any impression on agricultural production in the beneficiary communities. Much more needs to be done, therefore, to ensure that the local demand from the schools is benefiting local farmers. But pro-poor farming policies will need to better understand the nature of agriculture in Ghana, which is dominated by small-scale subsistence farmers, some of whom have as little as 1.6 hectares of land each, who account for about 80 per cent of national agricultural output. Although the sector has been growing in recent years, the development of agriculture in Ghana is stymied by a combination of inefficient farming practices, dependence on rain-fed agriculture and poor marketing outlets for farm produce. Consequently, some 58 per cent of food-crop farmers are classified as 'food-insecure' (WFP, 2006).

Calibrating demand and supply is proving to be a far more difficult task than anyone realized when the GSFP was initially designed. The Government of Ghana was not the only party to make unrealistic assumptions about the idea of creating local markets for local farmers: the international donor community was equally naïve in thinking that the home-grown model could provide 'quick wins' in the battle against hunger, while the experts who wrote the *Halving Hunger* report assumed that schools would provide a 'ready market' for the increased production of local farmers. In short, donors and agencies alike had proposed a hopelessly unrealistic timeline when they suggested that HGSF could be a quick and easy way to meet the MDGs (Morgan et al, 2007a).

Contrary to what some economists would argue, the forces of 'demand and supply' are not desiccated or impersonal laws of nature. In the case of the home-grown model, for example, 'demand' actually means the people who run the schools and the districts, who have multiple demands placed on them in addition to school feeding; while 'supply' consists of phenomenally poor subsistence farmers who need help, in the form of extension services, before they can increase production and come to understand that schools are something that they had never imagined them to be before – namely, markets in disguise. Procurement, as we saw in earlier chapters, can help to calibrate supply and demand, but it can only do so by recognizing the socially constructed and culturally embedded nature of these forces and relations. To overcome the procurement problems that have beset the GSFP, the donor community would need to drop the decidedly unhelpful language of 'quick wins' and 'ready markets' – expressions that neglect the fact that the home-grown model constitutes a steep learning curve for everyone involved.

The third problem that needs to be resolved as a matter of urgency is the funding of the GSFP. The Government of Ghana made a brave decision at the outset to fund its share of the programme through the national budget, which gives it a measure of continuity and stability. However, the current four-year programme is only viable because of donor support from the Dutch Government, which will end in 2010 (Fisher, 2007). How, if at all, will the programme be financed in the future?

The case for putting the GSFP on a sound financial footing rests on its intrinsic human value, both in and beyond Ghana. What needs to be better understood is that the GSFP is an important learning curve for the whole of Africa, a point that was acknowledged in October 2007, when politicians from 12 African countries met in Accra to draft a new road map for the NEPAD programme of HGSF. The executive director of the GSFP, Dr Kwame Amoako Tuffour, used the occasion to launch the idea of a new trust fund to which donors, NGOs and civil society groups could make financial contributions to sustain the programme (the costs of which, he said, were beyond the resources of the Government of Ghana). However, Roland Sibanda, the WFP liaison director to the OAU, stressed that governments should commit their own resources to the home-grown model before they sought funding from the international community. In our view, there is no real contradiction between these two perspectives, because, if it is to be a success, HGSF should ideally be funded by a combination of locally furnished resources (to demonstrate local commitment) and donor funds (to demonstrate global support and solidarity).

A good example of the local–global partnerships that will be required in the future is the practical assistance that the WFP is currently providing to raise the quality of the GSFP and extend its coverage, underlining the point that expertise is just as important as finance. For example, the basic WFP ration has been tested with the GSFP menus and adds significant nutritional value. To complement its contribution to consumption, the WFP is also helping to develop local production capacity by providing support to the private sector to produce and market the national fortified food delivery chains of iodized salt, palm oil, fortified corn–soy blend and maize meal. Practical partnerships like this need to be extended to include primary producer associations, which have the potential to deliver two major benefits:

1 to provide a means for small-scale farmers to share knowledge among themselves about things such as soil management, production techniques and routes to market; and
2 to help small-scale farmers to act in concert and aggregate their bargaining power vis-à-vis traders, processors and retailers in the food chain. (Morgan et al, 2007a)

The fate of the GSFP should be of concern to every developing country government because, despite appearances to the contrary, HGSF is about so much more than just school food. Among other things, it is about fashioning a robust and transparent framework for collective action. It is about creating, and sustaining, a dedicated budget to enable the system to survive the vicissitudes of the electoral cycle. It is about learning to deploy the power of purchase in a way that nurtures small-scale farmers, helping them through the transition from subsistence farming to commercial agriculture. And it is about keeping corruption at bay and enlisting the active support of civil society and the business community. To see the home-grown model in its true light is to see the whole in the part or, in the immortal words of William Blake, 'to see the world in a grain of sand'.

If this interpretation is correct, the home-grown model needs to be given a much higher political priority by developing country governments, because it has the potential to deliver multiple dividends, including health, education, agricultural development and female empowerment – all of which lie at the heart of the MDGs. Here the donor community has a major responsibility to help poor countries to help themselves, but this requires major changes in the way in which developed countries relate to, and engage with, developing countries. In short, the entire donor community needs to wean itself off the habit of confusing development with aid, a conflation of ends and means.

Learning by Doing: The Home-Grown Challenge

The home-grown model of school feeding is a more radical break with the past than even its architects realized, and therefore it can legitimately be called a *revolution* in school feeding. To be more precise, it is a revolution in the making, because designing a programme is one thing, delivering it quite another. Even in Ghana, the leading country in Africa in this regard, the implementation of the programme leaves much to be desired, especially with respect to the poor linkages between farms and schools, or, in other words, between production and consumption. As we have argued, this is as much a failure on the part of the donor community as it is of the Ghanaian Government; indeed, the former is more culpable because it ought to have known better than to suggest that local farmers could expect to find 'ready markets' for their produce in local schools. If the home-grown model is to be successfully implemented, a new partnership is needed between developed and developing countries in which *development* is understood in multidimensional terms – that is, as a concept that requires a holistic and integrated set of policies to be given practical meaning.

In the context of so many interrelated barriers to development, it might be thought that school feeding is too modest a programme to merit attention. Nothing could be further from the truth. When it is seen in its true light, the home-grown model is about the entire drama of development in microcosm. Learning to design and deliver a system of HGSF involves developing countries in a whole series of other learning curves – in public administration, financial planning, procurement, agricultural innovation and rural development, for example. The home-grown model, therefore, needs to be understood as a learning-by-doing exercise in which the end product, the provision of nutritious food, is just one part of a much larger process.

8

Sustainable Development and the Public Realm: The Power of the Public Plate

The welfare of the people is the highest law. (Cicero, *De legibus*)

Whatever the national context or cultural setting, the reform of school food raises some of the most compelling ethical, political and economic questions that a society can ask itself in the 21st century. Does the state have a duty to try to change the behaviour of its citizens for the better? Or does this constitute an unwarranted intrusion of the 'nanny state' into the private realm of personal freedom and individual choice? Can a society truly claim to be sustainable if it fails to invest in nutritious school food for young and vulnerable people? And when a society decides to invest, how can it convert its good intentions into good practice? In other words, what are the defining features of a sustainable school food service and are they applicable to both the poor countries of the South and the rich countries of the North? Finally, should societies seek to promote more 'localization' of their food and farming sectors in the name of sustainability or more 'globalization' in the name of fair trade?

To discuss these questions in a critical but constructive manner, we will draw on the concrete experiences of the different places that we have covered in this book, which provide an instructive benchmark for what the most innovative school food reformers have been able to achieve in the world of policy and practice. To complement this empirical focus, we will also draw on some theoretical debates in the social sciences. Although the issue of school food rarely figures in these debates, we believe it is a valuable prism through which to explore questions that exercise the social sciences today, not least the role of the public realm in fostering a more sustainable society and the ethical obligations that we, as human beings, owe to our 'nearest and dearest' as well as to 'distant others'. In short, this final chapter aims to address the theory as well as the practice of sustainable school food strategies in diverse cultural contexts.

A Public Ethic of Care: The New Moral Economy of School Food

The concept of the moral economy has re-emerged in recent years, partly as a response to the excessive utilitarianism of mainstream economics and partly as a vehicle to address normative issues that academics and activists consider to be *intrinsically* significant (like health and wellbeing). According to Andrew Sayer, one of the most prominent social theorists in this field, the moral economy:

> *embodies norms and sentiments regarding the responsibilities and rights of individuals and institutions with respect to others. These norms and sentiments go beyond matters of justice and equality, to conceptions of the good, for example regarding the needs and ends of economic activity. They might also be extended further to include the treatment of the environment.* (Sayer, 2000)

Although the concept can be applied to any sector of a modern economy (and we see pale imitations of it, perhaps, in today's corporate social responsibility policies), the school food service is surely the most germane context in which to think about a moral economy, because the consumers in question are *children*, who need protection, guidance and nurturing. In this sense, we argue that the moral economy of school food needs to be underwritten by, and integrated with, an *ethic of care*.

Broadly speaking, while the moral economy is largely associated with the principle of justice and with rights, an ethic of care has to do with relationships. As some feminist authors have argued, far from being formal and abstract, caring is tied to concrete circumstances, and it is best expressed not as a set of principles but as an activity. For feminists who champion an ethic of care, 'morality is not grounded in universal, abstract principles but in the daily experiences and moral problems of real people in their everyday lives' (Tronto, 1993, p79).

These feminist views are very pertinent to the moral economy of school food for two main reasons. On the one hand, they help us to understand why the central activity at the heart of the school food service – *caring* – has been devalued in the public realm of high politics. On the other hand, they also highlight the potential of the school food revolution in creating a new ethic of care that is more attuned to the principles of sustainable development. The pioneering work of Joan Tronto on the nature of care in capitalist societies provides an ideal starting point to develop this argument.

In *Moral Boundaries: A Political Argument for an Ethic of Care*, Tronto raises the central question of 'why care, which seems to be such a central part of human life, is treated as so marginal a part of existence' (Tronto, 1993, p111). In addressing this question, the book provides a trenchant critique of the claim that politics would be more moral if only more women were involved in it – a claim that is often made by feminists to advance the cause of women in public life. In particular, Tronto states that the 'women's morality' argument, which emphasizes a familiar set of appealing values (such as caring and nurturing, the importance of maternal love, the value of sustaining human relationships and the overriding value of peace), has been spectacularly unsuccessful because it failed to advance the cause of middle-class women in political life and marginalized poor and ethnic minority women. The author expresses this core argument in paradoxical terms by saying that 'we need to stop talking about "women's morality" and start talking instead about a care ethic that includes the values traditionally associated with women' (Tronto, 1993, p3).

What really derailed the 'women's morality' argument, according to Tronto, was the fact that an ideological boundary between the public and private realms had become established by the end of the 18th century. The public realm was a male-dominated sphere where reason and politics held sway, while the private realm was the sphere of sentiment and morality, to which women were confined. Many feminists inadvertently reinforced this ideological boundary, in Tronto's view, when they relegated caring to the private world of personal relationships, which is precisely what mainstream Western philosophy has done since the 18th century. The challenge for feminism, and indeed for progressive politics generally, is to develop an ethic of care that is not exclusively associated with the private realm.

To construct her argument, Tronto draws on a definition of care that she developed jointly with Berenice Fisher:

> *On the most general level, we suggest that caring be viewed as a species activity that includes everything that we do to maintain, continue and repair our 'world' so that we can live in it as well as possible. That world includes our bodies, our selves and our environment, all of which we seek to interweave in a complete, life-sustaining web.* (Fisher and Tronto, 1991, p40)

According to Tronto, several points emerge from this nuanced definition of caring. First, care is not restricted to human interaction: it also embraces objects and the environment in general. Second, there is no presumption that caring is dyadic or individualistic, so it is clearly not tethered to the most stereotyped caring relationship of all – the romanticized mother and child relationship. Third, the activity of caring is largely defined culturally and will therefore vary between different societies. And fourth, caring is not just a cerebral concern or a character trait, but the concern of living, active human beings engaged in the processes of everyday life.

Tronto's work helps us to explain two defining features of the school food service. First, it helps us to appreciate that the low status of the care-givers – that is, the largely female workforce that does the catering, the cooking, the serving and the caring – is a *systemic*, rather than a random, aspect of the service. Second, it helps to explain the low political priority accorded to the school food system for the better part of the 20th century. In simple terms, so long as care is largely confined to the private realm (supplemented by a minimal amount of care in the social services of the public sector), it will never be able to shed its Cinderella status in the worlds of work and politics. To overcome this problem, Tronto suggests that we need to think about care in a broader *public* framework: if it is to be created and sustained, 'an ethic of care relies upon a political commitment to value care and to reshape institutions to reflect that changed value' (Tronto, 1993, p178).

A *public* ethic of care of the kind Tronto suggests would help to elevate the status of care in our societies, giving it equivalence with the principle of justice. The moral economy requires an alliance of care and justice: without *care* the moral economy would lack compassion; without *justice* it would lack rights. Many theorists have come to the conclusion that care and justice should be treated as two sides of the same coin if we want to develop a morality with universal reach – in other words, one that offers practical support to 'distant others' as well as to 'nearest and dearest' (Clement, 1996; Smith, 1998; Held, 2005; Sayer, 2007).

As we saw in previous chapters, a rudimentary moral economy of school food emerged in the 20th century around the welfare state – largely, though not exclusively, in Europe. However, this old moral economy had three major shortcomings:

1 It was confined to very basic standards of care.
2 Its services were designed for, rather than with, citizens.
3 It had little or nothing to say about environmental stewardship.

Today, a *new* moral economy is beginning to emerge around school food that seeks to overcome the earlier shortcomings:

1 It is based on a much broader conception of care – enshrined, for example, in the 'whole-school approach' to healthy eating.
2 It offers far more opportunities to participate in the design and delivery of the service.
3 It shows more awareness of the need to embrace society *and* nature.

Collectively, these features also signal the emergence of a broader, more encompassing ethic of care. In the next two sections, we will explore the prospects for this new public ethic of care in two different contexts: first, the specific context of the school food system and second, the wider context of the political debate about food, health and children. Drawing on our case studies and on other examples, we will attempt to identify the various dimensions in which this emerging ethic of care needs to be strengthened and more consistently implemented if we want the new moral economy of school food to contribute to the goals of sustainable development.

From Good Intentions to Good Practice: Cultivating a Sustainable School Food System

More an ecology than a service, a sustainable school food *system* covers a much wider spectrum of activities than a conventional school meal service. In addition to dealing with the traditional aspects of the food chain, it also concerns itself with the environmental impact of the service – with the carbon footprint of its suppliers and the treatment of its own waste, for example. The more a school food service embraces these pre- and post-fork activities, the more it conforms to a sustainable school food system. To explore the contributions that the new public ethic of care can make to the delivery of sustainable school food systems, we will focus on four key dimensions:

1 the whole-school approach to healthy eating;
2 school catering;
3 food procurement; and
4 the supply chain.

In each case, we try to show the scope for, and the barriers to, a more sustainable school food system.

'I eat therefore I am': The whole-school approach to healthy eating

Far from being autonomous, hermetically sealed spaces, schools are both in and of their communities. Hence, they cannot be expected to solve societal problems on their own. For example, although it is beginning to take root in schools around the world, the 'healthy eating' message embedded in the whole-school approach has run into two major obstacles. First and foremost, it has been overwhelmed by the 'junk food' message, which dwarfs it in terms of advertising spend. And second, the public health community has erroneously assumed that getting the right information to the public would be sufficient to induce behavioural change, as though people behave in a purely rational manner when they make decisions about their personal welfare. Despite these obstacles, however, the whole-school approach can have a positive influence on what children eat, both in and beyond the school; therefore, it plays a key role in fostering the *demand* for healthier food in schools.

At its best, the whole-school approach works at a number of different levels. Besides aiming to change children's tastes, it also tries to embed the healthy-eating message into a wider educational package that stresses the positive links amongst food, fitness, health and both *physical* and *mental* wellbeing. In short, the healthy-eating message has to run through every aspect of the school – but especially the classroom, the dining room and the vending machine – to ensure that the landscape and the mindscape of the school are compatible and mutually reinforcing.

Parents and school caterers alike are painfully aware that children's tastes cannot be transformed overnight, especially where junk food is concerned. Quite apart from its cultural and political associations (where children can use it to express 'cool' or even 'deviant' behaviour), according to some researchers junk food is also profoundly addictive. If confirmed, this finding will totally undermine the standard food industry claim that food choice is free choice. Children and young people can be induced to change their eating habits so long as they are involved in the process. To illustrate this simple but fundamentally important point, we draw on two whole-school approaches from different ends of the food culture spectrum: those of Italy and the UK.

The most imaginative whole-school approach we encountered in our research was undoubtedly the Cultura che Nutre programme in Italy. As we showed in Chapter 4, the architects of this approach were remarkably candid about the social engineering dimension of the programme, which was designed to cultivate 'future consumers' for local and regional food and 'influential messengers' to their families about healthy diets. In this sense, Cultura che Nutre captures the essence of the Italian whole-school approach, which sees school meals as an educational tool to promote the values and meanings attached to food. It is crucial to emphasize here that in the Italian system these values and meanings are not seen as a passively inherited legacy but as something that must be created anew in each generation (Morgan and Sonnino, 2007, p22). Through initiatives such as Cultura che Nutre, Italian authorities actively intervene to continually remind future consumers of a very simple but powerful idea conveyed by the title of one of the programme's books: 'I eat, therefore I am.'

A creative whole-school approach can also capture the imagination of children in countries that do not enjoy such a rich food heritage as Italy, as Scottish reformers have understood. Another example comes from Wales, where a very poor community was chosen to pilot a highly ambitious food education project, called 'Food Matters', which was designed to nurture children's knowledge of the school food chain from 'seed to plate'. One of the most innovative features of the project was a peer-to-peer learning process through which children were taught to teach other children about the merits of healthy eating. Eating what they had grown, learning from each other and, above all, being actively involved in their own transformation were the hallmarks of the Food Matters project, which proved that children in the most deprived communities can learn to appreciate healthy food if the learning opportunities are sufficiently stimulating and empowering (Morgan et al, 2007b).

These examples demonstrate that a disposition for healthy eating is a socially acquired facility, rather than something inscribed in one's genes or history; it is the result, in short, of learning with family and friends, at home and at school. Where the whole-school approach is fun, stimulating and enabling, it can deliver handsome dividends even in the most challenging social environments. To make a school food system sustainable, one needs a steady demand for healthy food from young eaters in school canteens. If the whole-school approach can succeed in nurturing young people who genuinely understand the positive links between food, health and wellbeing, it will have fashioned the single most important ingredient in the recipe for a sustainable school food system.

Between commerce and health: School catering in transition

If healthy school food is vital to the wellbeing of children, as governments around the world now seem to realize, then the people who deliver that food – the school caterers – are effectively health workers in disguise. Although the status of school caterers is beginning to improve, as society becomes more conscious of their contribution to health and wellbeing, this is progress from a low base.

The school catering challenge varies considerably from one country to another, with developing countries facing the biggest challenges of all. Even in developed countries, however, school caterers sometimes have to work in very difficult operating environments. In the highly commercial environments of the US and the UK, caterers have to perform minor miracles by making enough money to support the service in a context of *choice*, where junk food is available alongside healthier options. Contrast this experience with Italy, where caterers do not have to worry about choice because there is none, or with Finland and Sweden, where free school meals are offered to all children and financed out of general taxation as a long-term investment in public health and social justice. Removing choice from the dining room, or making school food a fully funded part of the health service, means that school caterers in Italy, Finland and Sweden are not encumbered by the commercial pressures that daily confront their counterparts in the US and the UK.[1]

As earlier chapters have shown, the challenging environments in the US and the UK have not deterred school caterers from trying to design a more sustainable school food system. In addition to our featured case studies, we can point to other pioneering school caterers who are also working against the grain of their national food system to deliver a high-quality service. There is no better example of this in the American school system than Ann Cooper, a former celebrity chef and now a self-proclaimed 'renegade lunch lady'. As the director of nutrition services in Berkeley, California, she is spearheading the transformation of the school lunch programme, which was largely based on frozen food and is now largely based on fresh food. Where Berkeley schools used to serve fruit once a week, it is now on the menu every day along with a salad bar (Cooper and Holmes, 2006).[2] The UK also has transformative caterers, the most celebrated of whom is Jeanette Orrey, a dinner lady who decided to withdraw from her local authority contract in Nottinghamshire to prove that locally sourced school meals could be affordable and popular if school caterers jettisoned 'cheap processed muck' and reverted to cooking from scratch (Orrey, 2003).

Despite the success of these well-known caterers, the situation in countries like the US and the UK raises major challenges. In the UK, the vast majority of school caterers belong to the Local Authority Caterers Association (LACA), which has been struggling to rise above the commercial constraints of the service for many years. Even now that the service is receiving more recognition and new investment, LACA claims that the transition to a more sustainable catering service is stymied by the fact that school caterers are expected to deliver a *welfare* service whilst they are still treated as a *commercial* service (see Box 8.1).

Box 8.1 *LACA survey reveals major downturn in secondary school meals*

One of the key findings of a new national survey into school meals by the sector's leading representative body, the Local Authority Caterers Association (LACA) is that school meal numbers have dropped in over 75 per cent of local authorities.

One of the key factors that will be impacting on school meal prices and consequently, parents' pockets is that the costs involved with producing school meals have risen significantly. As a result of the new school meal standards requiring greater use of fresh ingredients cooked from scratch, the average food costs for primary school meals has gone up from 40p in 2004 to 60p in 2007, whilst in secondary schools it has increased from 56p in 2004 to 74p in 2007. With food requiring greater preparation time and more staff, labour costs for pay and training have also shot up. All of these factors have seen school meal prices soar.

Despite an investment from the Government to bring about the transformation agenda for school meals, most local authorities are struggling to cover the costs involved in making the improvements required by the new regulations. In 2004, nearly all respondents either broke even or made a small surplus which was ploughed back into the school catering service. However, in 2007 the picture is one of considerable concern over the future viability of the school meals service, particularly in secondary schools. 91 per cent of local authorities have reported that they are either breaking even (42 per cent) or in deficit (51 per cent). Few are now making any surplus at all.

Commenting on the findings of LACA's 2007 School Meals Survey in England, LACA Chairman, Sandra Russell says:

> *Our 2007 Survey findings confirm the feedback LACA has been receiving from its members over the past year and reflects our worst fears about the future viability of the school meals service in our secondary schools. We cannot expect to reverse and embedded eating culture overnight nor can we convert teenagers to a healthier regime by force. We are in danger of the secondary school meals service fragmenting or dying altogether if we are not careful.*
>
> *We must see common sense prevail. We, as school caterers, are being expected to provide a welfare service whilst still having to operate as a commercial venture. Whilst LACA and its members are totally supportive of the long-term aims of the new standards, we believe that the introduction of such radical change to young people's dietary habits is too draconian and too fast. The service is under immense pressure and already being seen by many private contractors as a non-viable operation. Our concern is that soon it may be for public sector caterers too.*

The LACA survey provides a brutally frank assessment of the problems encountered by school caterers as they try to make the difficult transition from a fully commercial service to a health and wellbeing service. As we mentioned in Chapter 5, LACA members are squeezed between two new pressures: their operating costs are increasing while their school meal take-up rates are falling. Because the higher costs of fresh food and extra labour time are here to stay – since they are a necessary feature of a sustainable school food system – the service could become financially unsustainable unless it can boost its take-up rates.[3]

The prospects for boosting take-up rates will depend on a whole series of factors, some of which are internal to the school, like the involvement of children and parents, while others are beyond its influence, like the resources that are made available to fund the transition and the complementary regulations that are designed to sustain it. Public service experts support this analysis when they argue that the quality of the food, which is the one thing that caterers can control, is not in itself sufficient to boost take-up rates. In the UK, for example, public service experts say that school management and environments, particularly the dining environment, can have a significant impact on service uptakes and financial stability (Baines and Bedwell, 2008, p5).

The fact that take-up rates vary so much within the same local authority suggests that school caterers, who are often blamed for poor take-up, are only one factor among many in the school food equation; in fact, head teachers, teachers, governors and parents often have more influence over the 'tone' of the school. What is perhaps most extraordinary is that, despite their relative powerlessness, school caterers are expected to be renaissance-like figures. The Chair of the School Food Trust in the UK could have been speaking for caterers everywhere when she said:

> *Caterers have to be sales and marketing people now, as well as craftsmen, nutritionists, people managers, accountants and social workers! School Cook, or Dinner Lady, or even Catering Manager, hardly describes the job.* (Leith, 2007)

All these skills will be needed if caterers are to boost the take-up of free school meals, considering that the eligible children are by definition from the poorest families in the community. Financial poverty can be compounded by nutritional poverty because, for a whole series of reasons, including the social stigma attached to free food, eligible children tend to recoil from their entitlement, as we saw in the case of New York.

Although it is certainly not confined to the US, the problem of getting poor children to eat free food does seem to be more difficult in the American context. The social stigma appears to be greater in the US because, quite apart from the cultural pressures to be independent, to 'stand on one's own feet' and so forth, the beneficiaries of the free lunch programme are physically segregated from 'regular' children who can pay their way. US regulations stipulate that foods of minimal nutritional value (in other words, junk food) must not be served in areas where the free lunch is delivered because the latter must meet certain nutritional standards. School caterers are set up to fail here because, no matter how skilled or committed they are, they cannot possibly counteract the negative impact of federal regulations that have the perverse effect of rendering social stigma more, rather than less, visible. It is no coincidence that the social stigma of free school food has been managed best where school caterers are allowed to offer the same food to payers and non-payers alike and where a cashless payment system makes it impossible for children to tell the difference between them. Where school caterers have the freedom and the resources, these are the innovations that they are trying to introduce to boost take-up, as we have seen both in the case of New York and in that of South Gloucestershire and Carmarthenshire. For school caterers, higher take-up serves a double purpose: it puts the service on a sustainable *economic* footing, which helps to secure their jobs, and it meets the *welfare* ethos of their service by ensuring that the poorest children have a nutritional meal.

Creative procurement: Buying into sustainability

One of the most important points that we have tried to establish throughout this book is that procurement, the most neglected tool in the public sector toolbox, needs to be taken far more seriously if countries want to create sustainable school food systems. In Chapter 2 we suggested that sustainable food procurement was essentially 'about defining best value in its broadest sense'. Unfortunately, the definition of 'best value' is more of an art than a science, so we should not be surprised by the lack of consensus around the definition. In cost-based contracting cultures such as those of the UK and the US, the biggest barrier to sustainable procurement has been a systemic tendency for low cost to masquerade as best value.

Like the UK, Italy is also subject to EU public procurement regulations, but it has interpreted these rules in a totally different way. Where the UK was conservative, Italy was bold; where the UK stressed value for money in the narrow economic sense, Italy sought values for money in the broadest sense of the term. The explanation for these divergent interpretations is to be found in the interplay of cultural values and political willpower, which in Italy's case sets a high premium on the creative procurement of produce that is strongly associated with seasonality and territoriality. Far from being the barriers they are often deemed to be, the EU procurement regulations are therefore more enabling and less prohibitive than many in the UK tend to think, but only if countries have the competence and the confidence to use the power of purchase to support more sustainable food chains.

If the European regulations have been a source of confusion as to what public bodies are allowed to do with respect to procurement, the US regulations pose no less of a problem. As we saw in Chapter 2, USDA interprets the regulations in a very conservative fashion, claiming that school districts are not allowed to specify local geographic preferences when they issue their tenders, an interpretation which is fiercely contested by other legal experts. Nothing will do more to promote the cause of local school food procurement in the US than a clarification of the federal regulations, so that local sourcing is positively encouraged rather than 'not disallowed'.

The US situation is even more perplexing at the state level because, whether from fear or ignorance, some states are not using the powers available to them under existing procurement regulations. A regional Farm-to-School conference in Michigan, for example, was shocked to discover that its own state had failed to make full use of the 'small purchase threshold', a less restrictive contracting procedure that is designed to encourage more local food procurement. The Michigan Board of Education was found to have capped this 'small purchase threshold' at just under US$20,000 per annum for each school district, when federal regulations allow a threshold of US$100,000. Even small school districts spend nearly US$2 million on food annually, so the Michigan state cap means that local procurement is limited to one per cent of their annual food budget. A school food service consultant captured the bemused mood:

> *Here we were, 330 of the region's professionals in school education, nutrition and health at the table trying to figure out how to nudge our regional food onto the centre of the plate in our schools' dining rooms – a gargantuan task in the commodity-driven, low-bid atmosphere of school food – and we find out our own state isn't setting a stage to help itself in this effort. Schools and prisons are the two largest state-run agencies that handle foodstuffs. Ironically, it's actually easier, due to lack of Federal regulation, to feed incarcerated citizens locally grown food than it is to feed it to our own children.* (Collins, 2008)

Where Michigan is a laggard, California and New York are leaders, since each of them has passed enabling state legislation to tap the potential of local food procurement. Many other states are moving in this direction, as they come to realize that where they were once deemed to be laboratories of democracy, they are now expected to become laboratories of sustainable development. While more supportive federal regulations would expedite this process, the crucial point to establish is that local food procurement is taking off in the US despite the national system, rather than because of it.

Supplying the schools: Short, long and sustainable food chains

New types of supply chain tend to emerge when schools buy into sustainability. Although these new supply chains set a high premium on the use of fresh, locally produced food, we must remember that *sustainable* food systems are not wholly synonymous with *local* food systems. As we showed in Chapter 4, the overriding feature of school food reform in the City of Rome was its emphasis on quality food rather than local food, though a more localized food chain remains part of its long-term goal. The stress on local food was also qualified in London, where the Mayor's strategy emphasized the importance of culturally appropriate food. In the context of a world city, this involves fairly traded food from afar as well as locally produced food from the city's regional hinterland. In both Rome and London, then, localization is deemed to be important, but not to the point where it overrides all other considerations.

Because of their local food tradition, cities and regions in Italy find it easier to craft more localized supply chains than those in the UK or the US, where mainstream suppliers tend to be part of large national and international food service companies. Despite their corporate muscle and their entrenched position, these companies have been chastened by the school food revolution, not least because the pressure for healthier menus has eroded their profit margins. Some low-quality meat suppliers in the UK have been forced out of business, and one of the very biggest suppliers, Compass-owned Scholarest, has withdrawn from a number of large contracts, including a county-wide contract for 172 schools in Bedfordshire and a 45-school contract in the London Borough of Richmond (Druce, 2007).

National food service companies should not be dismissed *en bloc* – in some cases, they are prepared to do more local sourcing if the produce is available, as has happened in South Gloucestershire. However, as the case of East Ayrshire in particular shows, there is no doubt that the demand for healthier school food creates opportunities for local economic development if local suppliers have both the appropriate produce and the infrastructure to distribute it.[4]

Securing local food from local suppliers is one of the hallmarks of the Farm-to-School movement in the US, where over 1000 programmes in 38 states are now buying fresh products from local farms. This movement has the potential to become one of the most important social movements in the country, as it fuels the growth of more sustainable food chains and helps communities, through their schools, to reconnect to the producers of their food. According to Marion Kalb, the Director of the Community Food Security Coalition's Farm-to-School Program, what would boost Farm-to-School more than anything else would be a reform of the federal meal reimbursement rate to give school districts a greater financial incentive to use fresh, locally produced food rather than the commodity foods that are privileged today.[5]

A backlash against commodity foods could be triggered by the truly shocking state of animal welfare at the Hallmark/Westland Meat Packing Company plant in Chino, California, which supplied a fifth of all the ground beef used by USDA for nutrition programmes, including the school lunch programme. The Humane Society of the United States secretly filmed the ghastly conditions at the plant and showed it to a worldwide audience on YouTube. The film included the chilling sight of downer cattle (cattle that are unable to walk) being forced to their feet before being slaughtered for ground beef, despite the fact that USDA had banned such meat from the school lunch programme in 2000. Although the company recalled some 143 million pounds of ground meat on 17 February 2008, the biggest food recall in US history, the incident fuelled fears about conditions at other commodity meat plants.

The beef recall raises serious questions about the role, or roles, of USDA, which has been less than vigilant in regulating the commodity food chain. In reports dating back to 2003, the USDA Office of Inspector General and the Government Accountability Office cited USDA's lunch programme administrators and inspectors for having weak food-safety standards and poor safeguards against bacterial contamination and choosing lunch programme suppliers with known food-safety violations (Williamson, 2008). Far from being fostered by federal regulations, sustainable school food systems in the US have been stymied by USDA in at least two ways: first, through its unwavering commitment to the commodity food chain, which trades in cheap food of dubious provenance; and second, through its deeply conservative interpretation of federal regulations, which it construes to mean that states and school districts are not allowed to purchase fresh, locally produced food. The USDA system provides a perfect illustration of the way in which *national* regulations are used to circumscribe *local* initiative. As well as lacking transparency, this system also contains enormous conflicts of interest. In fact, in addition to its responsibility for the national school lunch programme, USDA is also responsible for inspecting and marketing the commodity food system. In other countries, like the UK for example, these tasks have been separated because it was found that government departments that sponsor producer interests cannot be credible champions of the consumer – especially when the consumers are children (Morgan et al, 2006).

Each of the elements considered here – a whole-school approach to healthy eating, a health-promoting catering service, creative public procurement and a sustainable food supply chain – raises its own challenges. The real challenge, however, is how to synchronize these elements in time and space so that they have a mutually reinforcing, synergistic effect. The islands of good practice that we have documented in this book prove that real progress is possible as things stand today, but this is very much a case of progress being achieved despite, rather than because of, the current regulatory framework, which in most countries fails to take either sustainability or child welfare as seriously as it should. The emerging ethic of care clearly needs to embed itself into new regulations that are designed to make good practice the norm, rather than the exception.

Children's Diets and the Public Realm: The Crises of Obesity and Hunger

In both developed and developing countries, the school food service is now asked to address societal problems that are way beyond its traditional remit. In the rich countries of the North, it is expected to help redress the epidemic of childhood obesity, while in the poor countries of the South it is expected to help meet the MDGs, as we discussed in Chapter 7. A sustainable school food service can make an important contribution to these wider goals and, more generally, to the three fundamental principles of sustainable development. But to achieve these objectives, the school food service requires far more support from the public realm, and particularly from the state, which is the political expression thereof. Among other things, a more supportive public realm would have at its heart a strong ethic of care for citizens in general and for children in particular. After exploring the prospects for a public ethic of care in the context of sustainable school food chains, we now turn to two problems that threaten the achievement of the goals of sustainable development in all societies: the burgeoning problem of childhood obesity and the deepening problem of hunger. In our view, a public ethic of care is necessary to address this 'double burden of malnutrition' (SCN, 2006), as many experts call it.

Childhood obesity and the public ethic of care

No public health problem in recent years has commanded as much worldwide media attention as obesity. In the measured words of the World Health Organization (WHO), obesity 'has reached epidemic proportions globally, with more than 1 billion adults overweight, at least 300 million of them clinically obese, and is a major contributor to the global burden of chronic disease and disability'. One of the most surprising and counterintuitive aspects of this epidemic is the fact that it is a *global* problem. In 2007, for example, an estimated 22 million children under the age of five were overweight throughout the world, and more than 75 per cent of them lived in low- and middle-income countries.

According to the WHO,[6] global increases in childhood overweight and obesity are attributable to a number of factors, including (a) a global *shift in diet* towards increased intake of energy-dense foods that are high in fat and sugars but low in vitamins, minerals and other healthy micronutrients; and (b) a trend towards *decreased physical activity levels*, due to the increasingly sedentary nature of many forms of recreation, changing modes of transport and increasing urbanization. But childhood obesity is also a product of public policies in agriculture, transport, urban planning, the environment, food processing, distribution and marketing, as well as education. In short, according to the WHO:

> *The problem is societal and therefore it demands a population-based, multi-sectoral, multi-disciplinary and culturally relevant approach. Unlike most adults, children and adolescents cannot choose the environment in which they live or the food they eat. They also have a limited ability to understand the long-term consequences of their behaviour. They therefore require special attention when fighting the obesity epidemic.*

Curbing childhood obesity, according to the WHO, requires sustained political commitment and the collaboration of public and private stakeholders, including governments, international aid organizations, NGOs and the private sector, all of whom are asked to contribute to the Global Strategy on Diet, Physical Activity and Health – the worldwide strategy that the WHO adopted in May 2004.

The great merit of the WHO's analysis is that it stresses the *societal* nature of the problem, which demands a *population-based* solution. It is wholly inappropriate, then, to attribute the problem to feckless behaviour on the part of children and adults or to address solutions to people as individuals, like exhorting them to show more 'willpower' in their eating habits. These individualized prescriptions, which usually make little or no impression, continue to be canvassed for one very simple reason: they provide a source of profit for companies selling individual solutions – like weight-loss medicines, diet plans and, for the morbidly obese, gastric surgery. Although individuals are ultimately responsible for what they eat and how they choose to live their lives, the clear implication of the WHO analysis is that governments have a duty to take the lead in designing societal solutions to a societal problem. And, more generally, the public realm has a duty to debate these problems in a reasoned and transparent manner.

From a theoretical standpoint, this argument can be situated in the early work of Jurgen Habermas (1989), one of the leading social theorists of the 20th century. For Habermas, the public sphere as it emerged in 18th-century Europe was an independent space, situated between civil society and the state, in which individuals came together on a voluntary basis to confront the issues of the day and to arrive at some approximate notion of the common good. Although the bourgeois public sphere was nominally open to all, in reality it was confined to the educated and wealthy members of society, rendering it more partial than it appeared. This early public sphere only retained its critical and autonomous character until the 19th and 20th centuries, when its positive features were eroded by the advent of big business and the bureaucratic state, which managed to manipulate public opinion, converting citizens into passive and uncritical consumers. Despite this extremely negative trend, however, Habermas suggests that the public sphere is not destined to atrophy and decline; on the contrary, it could be deepened and broadened if reasoned debate and democratic renewal are forthcoming.

Whatever its limitations, the concept of the public sphere provides a salutary reminder of what really constitutes a democratic polity. For Habermas, a public sphere worthy of the name is constituted by two fundamental things: the quality of critical discourse and the quality of popular participation (Calhoun, 1992). Where they are present, these two qualities can help to foster a more robust political debate and ensure that the public realm, particularly the state itself, is not colonized by, and subordinated to, private interests.

Colonization and subordination are precisely what some critics claim have been under way in the public sector of many countries in recent years. In the UK, for example, some political scientists have argued that the state was subjected to a relentless *Kulturkampf* in the 1980s and 1990s to root out the traditional culture of service and citizenship, blurring the distinction between the public and private spheres (Marquand, 2004). Clearly, it is the duty of the state, as the expression of the public realm, to set the 'rules of the game' by which business conducts itself. Corporate responsibility for children's diets is also important, of course, but capitalist societies are still 'poorly equipped to cope with children being consumers' (MacMillan et al, 2004). Continuing ambiguity about the child as consumer makes it all the more important for this relationship to be subject to a clear public ethic of care.

Recent history suggests that governments enter this territory at their peril, especially when they seek to reform the food choice environment – one of the most ideologically charged issues in modern politics, because it pits the individual's freedom of choice against the state's duty to promote what it construes to be the common good. When confronted by a reforming government, the food industry draws on its well-rehearsed repertoire, which is to argue that food choice is a private matter, not a public one. Governments that try to promote a more healthful food environment are invariably accused of behaving like a 'nanny state' or in a 'totalitarian' fashion, because they are deemed to have crossed a moral boundary and made an unwarranted intrusion into the private realm of the individual (Nestle, 2002; Morgan et al, 2006).

Reforming the food choice environment may be just one part of an anti-obesity strategy, but it is the most incendiary one. Indeed, in political terms the campaign for a more healthful food system goes to the very heart of today's 'food war', a conflict that has been aptly described as a 'global battle for minds, mouths and markets' (Lang and Heasman, 2004). To illustrate the controversial nature of food choice, let us briefly examine the campaign to restrict the marketing of unhealthy food to children – perhaps the best index of a government's willingness to confront the junk food message.

The marketing of junk food – food and drink that are high in calories but low in nutritional value – is nowhere as developed as it is in the US, the home of the fast-food diet and of the most obese nation on Earth. Advertisers are way ahead of the Government in the US when it comes to getting the message across. During the course of a year, the average American child is thought to watch more than 40,000 TV commercials, half of which are for junk food. US fast-food chains spend more than US$3 billion annually on TV advertising, and this does not include their rapidly growing outlays on new media adverts or toys, which are said to be the key to attracting children (Schlosser, 2006).

From an ethical perspective, what is most shocking about the ubiquitous junk food message in the US is the fact that it seems to acknowledge no moral boundaries; in other words, it has little or no respect for children's spaces or for consumers who are too young to exercise their critical faculties. Most perverse of all is the fact that schools, where a duty of care might have been expected to nurture safe havens for vulnerable people, have been allowed to become spaces where junk food consumption is actively encouraged.

No one has done more to highlight the corporate reach and influence of the food industry on food, health and nutrition in the US than Marion Nestle. In her seminal book *Food Politics*, Nestle argues that the most egregious example of food company marketing practices is the deliberate use of young children as sales targets and the conversion of schools into vehicles for selling junk foods. The practice of 'pouring rights' we described in Chapter 3 perfectly captures this ethically challenged behaviour. As Nestle explains:

> *These contracts usually involve large lump-sum payments to school districts and additional payments over 5–10 years in return for exclusive sales of one company's products in vending machines and at all school events. [...] The most questionable aspect of these contracts is that they link returns to the companies and to the schools to the amounts that students drink.* (Nestle, 2002, p202)

Although some school districts have outlawed this egregious practice, either because they deemed it unethical or because they feared anti-obesity litigation, there is no national framework to regulate junk food in schools. At a time when many countries are banning junk food sales in the school environment, the US continues to exhibit an astonishing insouciance about food marketing to children. What other country, for example, would allow good school grades and attendance to be rewarded with a Happy Meal from McDonald's? Yet this is exactly what has happened in Florida, where the Seminole County School Board has signed a new deal with that company, replacing a ten-year-old agreement with Pizza Hut. Not surprisingly, this 'report card incentive scheme', as it is called, has caused an outcry from health and advocacy groups. 'Turning report cards into ads for McDonald's undermines parents' efforts to encourage healthy eating', said Susan Linn, director of the Campaign for a Commercial-Free Childhood (Elliot, 2007).

Such is the intensity and pervasiveness of the junk food message in the US that it amounts to a de facto masterclass on how food companies can successfully 'market obesity to children' (CSPI, 2003). Food companies will not willingly forgo marketing to children in schools. As the Center for Science in the Public Interest (CSPI) has shown, these educational spaces are highly valuable commercial sites, not least because children constitute a captive audience in school, where lifelong brand loyalties can be established early. Although school districts feel financially beholden to the junk food system, the latter is sustained by a whole series of myths – like the myth that junk food is the only lucrative way to raise funds for schools, the myth that junk food is the only food that children will buy and the myth that vending machines raise a lot of money for schools (CSPI, 2006 and 2007).

Because the junk food message is so ubiquitous, concerned Americans might well wonder where to begin to reform their food choice environment. The answer, as Marion Nestle says, is quite clear:

> *One place to begin is with children. If the roots of obesity are in childhood, then the marketing of foods to children deserves substantial public opposition. Banning commercials for foods of minimal nutritional value from children's television programmes and from schools, and preventing such foods from replacing more nutritious foods in school lunches, are actions ripe for advocacy – school by school, district by district, state by state.* (Nestle, 2002, p370)

Some American school food reformers hoped that the National School Nutrition Standards Amendment, sponsored by the Democratic Senator Tom Harkin (from Iowa) and the Republican Senator Lisa Murkowski (from Alaska), would be the first federal legislation to update the nutrition standards after 30 years. However, this amendment to the Farm Bill, which would have banned junk foods in school cafeterias and vending machines, was dropped in the Senate for two different political reasons: Democrats objected to federal pre-emption of stricter state standards, while Republicans were concerned about restrictions on snack foods (Black, 2007).

The failure to ban foods of low nutritional value from school premises underlines the fact that the US is unable or unwilling to confront the power of the junk food industry, leaving it with a woefully inadequate federal strategy to combat the world's worst obesity epidemic. Policies that could make a real difference, like creating a junk-free school environment, are deemed to be beyond the realm of feasible politics, while policies that are thought to be feasible, like education and information campaigns, have little or no public health impact (Brescoll et al, 2008)

A very different attitude is emerging in Canada, where a citizens' panel on food marketing to children issued a very clear consensus statement in 2008: 'Access to children is a privilege, not a right, and as such should be subject to stringent regulation' (CDPAC, 2008).[7] This is the simple, but fundamental, point that escapes the federal public realm in the US, which has signally failed to protect its children from unhealthy-food marketing campaigns in schools. And having failed to contain the junk food message to children, the US will have to shoulder the burgeoning costs of its obesogenic environment. Leaving aside the human costs, according to the US Surgeon General, the financial costs of obesity and obesity-related illnesses had escalated to US$117 billion per year in 2000 (United States Department of Health and Human Services, 2001). At some point federal politicians will have to weigh up the costs of obesity against the benefits of allowing the food industry to dictate the rules that govern the American food choice environment.[8]

Despite the superficial similarities with the US, there is a stronger political determination to tackle the junk food message in the UK, the country that has the highest level of childhood obesity in Europe. The UK Government is beginning to take a strategic view of the obesity issue following the Foresight Report, which shocked the British political establishment. By 2050, the report said, 60 per cent of adult men, 50 per cent of women and about 25 per cent of children under 16 could be obese and, without remedial measures, obesity-related diseases would cost society nearly £50 billion per year. The report emphasized that:

> *a bold whole-system approach is critical – from production and promotion of healthy diets to redesigning the built environment to promote walking, together with wider cultural changes to shift societal values concerning food and activity.* (Foresight, 2007)

The Foresight Report concluded that it will take at least another 30 years to tackle obesity. To meet the challenge of the Foresight Report, the UK Government launched a bold strategy to support the creation of a healthy society. As the Health Secretary explained:

> *It is not the Government's role to hector or lecture people, but we do have a duty to support them in leading healthier lifestyles. This will only succeed if the problem is recognized, owned and addressed in every part of society.* (Department of Health, 2008)

Although this 'cross-government strategy on obesity' signalled a promising start, it concealed a fierce behind-the-scenes political battle between the health and culture departments, with the latter winning the early skirmish to keep tougher rules on junk food marketing out of the initial version. This is now the big issue in the anti-obesity debate in the UK, where the Children's Food Campaign (CFC) has attracted the support of 300 organizations. One of the key aims of the CFC is to pass legislation that would:

- introduce a 9.00pm watershed for television advertising of unhealthy food; and
- protect children from other methods of marketing of unhealthy food.

A TV ban on junk food advertising was not considered to be sufficient on its own, since a consumer rights body revealed that food companies in the UK, like their counterparts in the US, were using irresponsible marketing ploys (including text messaging tricks, websites, games and free toys) to get their message to young children (*Which?*, 2006).

In an open letter asking MPs to support the Food Products (Marketing to Children) Bill, the CFC concludes that:

> *There is no one 'silver bullet' that will improve children's diets overnight. Making exercise a greater part of children's lives is important, as is parental responsibility. However, more effective regulation of the marketing of unhealthy food will play a key part in changing our food culture and improving our children's diets. The proposals in this Bill are a vital part of any solution to solve the obesity crisis.* (CFC, 2008)

The UK campaign is part of an international movement to reduce junk food marketing to children around the world. A federation of 50 national consumer organizations wants the WHO to adopt a tougher code to restrict junk food marketing, including outlawing the use of cartoon characters, celebrity tie-ins and free gifts aimed at children, all of which were part of the US$13 billion that the food industry spent on advertising in 2006 (Smithers, 2008).

Junk food advertisers contest the need for a ban on the grounds that TV advertising has only a 'modest' effect on children's food choices, but this argument totally misses the point about the obesity epidemic. No *single* action has anything other than a modest effect, because there is no 'silver bullet' or 'big idea' in the battle against obesity. Since this is a societal problem with multiple causes, every sector needs to be part of the solution. As for the campaign to reduce junk food marketing to children, the WHO is surely right to conclude that real progress will only be achieved when *public health* is placed at the centre of the debate (WHO, 2004). Societies that place public health at the centre of their deliberations are societies that have recognized the need for a new public ethic of care.

Hunger: The forgotten Millennium Development Goal

The idea that the developed world is grappling with obesity while the developing world is struggling with hunger fails to do justice to the nuances of food and health in the world today. While obesity is most prevalent among the lower socioeconomic classes in developed countries, the opposite is true in developing countries, where only the rich can afford to get fat (Lang and Heasman, 2004). Prosperity and poverty can live cheek-by-jowl in every country, but it is in the fastest growing countries of the developing South, where a Western diet is perceived to be part of modernity, that we see the starkest contrasts. In India, for example, the combination of urban growth and rural stagnation helps to explain the extraordinary coexistence of obesity and hunger – 'the double burden of malnutrition'.

Without wishing to downplay the obesity problem in the developing world, however, our main aim here is to highlight the need for a new anti-hunger campaign in the light of two potentially catastrophic trends:

1 the dramatic increase in world food prices for basic commodities such as corn, wheat, rice and soy beans; and
2 the mounting evidence that the MDGs will not be met.

These twin challenges will be a test of whether the international community has the collective resolve to demonstrate a stronger ethic of care for the poorest of the poor.

The recent dramatic increase in world food prices began in the latter part of 2007 and was driven by a new combination of factors, particularly the steep rise in oil and energy prices, the advent of biofuels, growing demand from China and India, and climate change effects (von Braun, 2008). The immediate effect of higher prices was social unrest throughout the developing world, involving sections of society, such as middle-class professionals, which had not previously been threatened by hunger. This new face of hunger, where high prices rather than food shortages are the main problem, has prompted governments everywhere to place *food security* at the top of their political agendas.

Burgeoning food and fuel costs forced the WFP to issue an 'extraordinary emergency appeal' to its donors to address a rapidly growing funding gap. The appeal (extracts from which are shown in Box 8.2) summarized the new challenge of hunger and called on donors to express a new sense of solidarity – a new ethic of care, in effect – to help the agency to rise to this challenge.

Under its new Executive Director, Josette Sheeran, the WFP has set the pace among the UN agencies on good governance, ethics and partnership working, all of which helps to assure the donor community that its donations are well spent. The agency is also becoming a better communicator, particularly in helping donors to appreciate what it would actually take to alleviate hunger at school. Addressing the Executive Board of the WFP, Sheeran said that 'in order for the world to say that no child goes to school hungry, it would cost about US$3 billion a year – in Africa alone about US$1.2 billion' (Sheeran, 2008).[9]

The international community now stands accused of failing to meet its MDGs (ActionAid, 2008).[10] As we mentioned in Chapter 7, the first MDG has two targets: to halve, by 2015 as compared with 1990, the proportion of people living on less than US$1 a day and that of people who suffer from hunger. Some real progress seems to have been made with respect to the first target – 980 million are estimated to currently live on less than US$1 a day, down from 1.25 billion in 1990. However, the international community is way off target on halving hunger. In fact, even before the food price explosion of 2007–2008, the

Box 8.2 *WFP letter of appeal to government donors to address critical funding gap*

The World Food Programme is today issuing an extraordinary emergency appeal to address the critical funding gap in our programmes created by soaring food and fuel prices. We urge your Government to be as generous as possible in helping us to close this gap – which stood at US$500 million on 25 February and has been growing daily. [...]

The price of food and fuel has soared to record levels in recent years, and entered an aggressive pace of increase in June of last year. The WFP has taken many steps to mitigate these increases, including making 80 per cent of our food purchases – US$612 million – in local and regional markets of the developing world. In 2007 alone, we increased our local purchases by 30 per cent. This not only saves on food and transport costs but is a win for local farmers, helping to break the cycle of hunger at its root. [...]

Of particular concern is the emergence of a new face of hunger. As stated by the UN Secretary General, Ban Ki-moon:

> *This is the new face of hunger, increasingly affecting communities that had previously been protected. Inevitably, it is the 'bottom billion' who are hit hardest: people living on one dollar a day or less. When people are that poor and inflation erodes their meagre earnings, they generally do one of two things: they buy less food, or they buy cheaper, less nutritious food. The result is the same – more hunger and less chance of a healthy future. [...]*

We urge your Government to act quickly on this request so that we may avoid cutting the rations to those who rely on the world to stand by them during times of abject need. We thank you for your steadfast support year in and year out, and urge you to stand by us as we rise to this challenge.

Josette Sheeran
WFP Executive Director
20 March 2008

number of hungry people in developing countries had decreased by just 3 million, to 820 million (FAO, 2006).

Prominent figures in the international community have freely conceded that they have a mounting crisis on their hands if they cannot get back on schedule. Speaking at the UN in 2007, UK Prime Minister Gordon Brown issued a 'call to action' to the international community to confront the fact that it was not on track to meet its own goals, precipitating what he called a growing 'development emergency'. The following year, when the world's business leaders made their annual pilgrimage to Davos, the President of the World Bank, Robert Zoellick, urged them to put hunger and malnutrition at the top of their agenda, calling the first MDG the 'forgotten MDG'. These rallying cries are perhaps the most palpable signs that the international community is anxious that, once again, it will fail to honour the most basic of all human rights – a tradition that, as we discussed in Chapter 7, it has upheld for 60 years.

One of the reasons why progress has been so dismal, according to ActionAid, is because the international community has failed to recognize, and act upon, the systemic discrimination against women and girls. As well as being a matter of social justice, gender equality is central to the entire development process, because women play a disproportionate role in reducing hunger and poverty – not least by improving family nutrition. All the evidence points to the 'development emergency' being first and foremost an emergency for women and girls, who, as the ActionAid report demonstrates in vast detail, are most likely to be poor, hungry, illiterate and sick.[11]

Greater gender equality helps to explain why some countries (Bangladesh, India, Mozambique and Tanzania, for example) have made rapid progress with girls' education. These countries have enjoyed strong political support for girls' schooling, allied to backing from high-profile women and civil society. Equally important has been the fact that girls' education was linked to a wider struggle to overcome discrimination through legal and social change. In short, ActionAid argues that more gender equality could help to put the MDGs back on track if urgent action is taken to:

- set more ambitious and specific targets on women and girls within the MDG framework;
- bolster the UN's capacity to tackle discrimination against women;
- monitor progress with better data; and
- make aid a more effective tool in achieving equality and women's empowerment. (ActionAid, 2008, p3)

Despite the 'little victories' that have been achieved, however, it is difficult to be anything other than pessimistic about the international community's chances of meeting the goals that it set itself in 2000, especially the goal of halving hunger by 2015. The momentum was lost even before commodity prices exploded in 2007–2008, since when each country has been trying to digest the food security implications for itself. When the coming food landscape is factored into the equation – with its higher prices and the uneven effects of climate change – the poorest of the poor will be hit hardest, particularly in sub-Saharan Africa (Ahmed et al, 2007). This new food landscape will contain one feature that is actually so old that we are in danger of becoming inured to it. As described by one of the world's foremost food policy experts:

> *The fact that large numbers of people continue to live in intransigent poverty and hunger in an increasingly wealthy global economy is the major ethical, economic and public health challenge of our time.* (von Braun, 2007)

In conclusion, the MDGs offer a tolerable definition of what sustainable development means in the context of poor countries, the bare minimum of what we should expect from an international community that prides itself on its civilized norms. These goals, especially the goal of eradicating extreme poverty and hunger, will be the quintessential test of the international community's commitment to sustainable development – a test, in other words, of its capacity to develop an ethic of care for the poorest of the poor.

Greening the State: From School Food Provisioning to Community Food Planning

'Greening the state', as we noted in Chapter 1, can be understood in two different ways. In the narrow sense, it merely connotes a state that is becoming more conscious of its environmental obligations, which is how most states understand the phrase in their everyday activities. This narrow conception of sustainable development is justified on the grounds that the environmental dimension is easier to measure and manage, compared to the social and economic dimensions. It is also the case, however, that the environmental side is perceived, by governments and corporations alike, as less threatening and more easily contained than issues of social justice and economic democracy, which constitute more of a challenge to the status quo. In a broader sense, however, greening the state refers to a more ambitious project, in which the state accords equal weight to all three dimensions of sustainable development and seeks to implement sustainable practices not merely in the public sector, but also, through the powers at its disposal, in the private sector (Morgan, 2007c).

This final section examines the scope for greening the state using the broader sense of the term. It focuses on two questions:

1. What can the school food experience bring to the Green State debate?
2. What are the prospects for using school food reform as a catalyst for community food planning?

School food and the Green State debate

We have talked rather casually about the rise of a Green State, but this idea is actually a highly contested matter in environmental politics. Indeed, the very possibility of a Green State under capitalist conditions is the subject of a lively debate between ecologically minded Marxists and ecological modernizers (Hajer, 1995; Hay, 1996). While eco-Marxists tend to see 'sustainable capitalism' as something of an oxymoron, ecological modernizers contend that the prospects for 'green growth' may be better than the critics allow, especially in countries, like Germany and the Nordic countries, where there is a strong ecological coalition that will pressure the state to exercise its powers to underline, rather than undermine, sustainable development. Context, in other words, is a critically important factor when assessing the prospects for greening the state.

The vast majority of the contributors to the Green State debate tend to be highly critical of the state-centric perspective, a view that essentially privileges the central institutions of the state over all other actors, be they civil society or corporate. In rejecting such a perspective, some writers go so far as to suggest that the state is actually neither necessary nor desirable from a green perspective because, being in thrall to the economic imperatives of capital accumulation or military competition, it is inherently bent on ecological degradation (Paterson, 2000).

A more judicious and realist perspective is that of Eckersley, who rejects the limiting state-centric view without rejecting the state itself. Writing from a critical political ecology perspective, she takes the green rejectionist lobby to task, reminding them that:

> *implicit in the day-to-day policy demands made of the state (both domestically and internationally) by environmentalists is a notion of what the state ought to be doing (or not doing) – in short, a green ideal or vision of what a 'good state' might look like.* (Eckersley, 2005, p160)

Far from being something to be avoided or suppressed, such normative theorizing actually creates new imaginaries as to what is feasible or desirable in the world; it helps to make hope practical, as the cultural theorist Raymond Williams was fond of saying.

Another merit of Eckersley's perspective is that it is fully alive to the continuing significance of the nation-state. Although she concedes that the autonomy of the nation-state has been constrained by globalization, she argues that there is a whole series of ways in which the state can help to foster a more socially progressive and sustainable society. To Eckersley (2005, p172), 'the appeal of the state is that it stands as the most overarching source of authority within modern, plural societies'. This analysis resonates deeply with our own research on public food provisioning, which has led us to totally reject the fashionable notion that the state has been reduced to a powerless victim of circumstance by the forces of globalization. Though the scope for unilateral action is constrained in manifold ways, the growth of multilateral institutions offers scope, however modest, for the state to act in concert with others to make a difference at the international level.

However, the problem-solving capacity of the modern state has internal as well as external constraints. Our research suggests that the state's *organizational capacity* – its capacity to regulate the economy, deliver public services and procure goods and services – needs to be better understood if we are to arrive at a finer appreciation of the scope for, and the limits to, sustainable development in capitalist societies. The Green State debate tends to focus on the external constraints to greening the state, not least the problem of regulating the protean corporate sector and persuading it to operate in a more sustainable fashion. However, a problem that is overlooked here is that the state would be in a stronger position to promote sustainable development in the private sector if it had already done so in the public sector – if it had put its own house in order, as it were.

What the school food reform experience reveals, albeit in microcosm, is that the values, metrics, skills and governance structures of the state tend to frustrate, rather than foster, its capacity to promote sustainable development across the public sector. Greening its own activities would be a steep learning-by-doing exercise, enabling the state to be a more credible, more talented interlocutor in its dealings with the corporate sector (Morgan, 2007c). Since organizational capacity is a key issue, especially to the Green State debate, it may be worth trying to distil some lessons from the school food reform experience.

The value of values

In cost-based public sector catering cultures, like those of the UK and the US, the school food service has been obliged to forage in the lowest-quality food chains, where food provenance leaves much to be desired. All public sector bodies are required to secure 'value for money', but, as we have explained, this can be interpreted in different ways. In cost-based contracting cultures, as we have seen, there is a strong tendency for low cost to masquerade as best value. Real 'value for money' needs to be buttressed by a broader, more sustainable metric, one that reflects a range of social and cultural values, rather than a single and narrowly defined economic value. It is difficult (if not impossible) for public bodies to harness the power of purchase to the cause of sustainable development if they know they will be evaluated, by auditors and lawyers alike, on the basis of a narrow commercial metric. Our case study of Carmarthenshire provides a sobering reminder that a high-quality school food service was nearly undone by a metric that extolled cost above all other values. Fortunately, Carmarthenshire was able to retain its high-quality service, but only after the value of other values (like public health) was recognized and brought into play. Politically, the most important point to establish here is that the values conducive to sustainable development – public health, democracy, environmental integrity and the like – do not appear automatically: they have to be actively fashioned and mobilized.

The power of purchase

Perhaps the most important single lesson from school food reform around the world is that procurement matters. What emerges from all our case studies is that public bodies everywhere are struggling to master the art of creative public procurement, where part of the challenge is to factor into the equation the costs – on health and the environment, for example – that are externalized in conventional accounting systems. Creative procurement is easier to implement where there is a politically supportive milieu, where public sector managers are not stymied by a narrow commercial metric and where they have the skills to apply whole-life costing methods – one of the keys to securing real value for money. No one should ever underestimate the powerful forces that are arrayed against creative procurement, not least the seven barriers that were identified in Chapter 2.

One of the most common barriers identified concerned 'legal issues', which refers to the uncertainty as to what can and cannot be done under existing rules on public procurement. Although these rules are gradually becoming more conducive to creative procurement, especially in the EU and in some states in the US, the majority of public bodies continue to act in a highly conservative, risk-averse manner when they design their tenders. What tends to be forgotten is that this risk-averse culture is assiduously cultivated, by the legal profession and management consultants, so it is not simply a reflection of a timid public sector ethos. As we can see in Figure 8.1, an advert for a training course in the UK cleverly plays on the fears of public procurement managers about failing to comply with the latest regulations. The ad went on to say that 'legal knowledge of the intricacies of the public procurement regime [...] is crucial'.

To overcome this risk-averse culture, and the fear factor that fuels it, the public procurement profession can help itself in two ways:

1. by becoming better organized as a public sector profession, allowing it to spread good practice through its ranks more effectively; and
2. by reaching out to other professions, particularly in health, economic development, environmental planning and transport, to broadcast the wider benefits of creative procurement.

As the public sector becomes more aware of, and attuned to, the power of purchase, the procurement profession could find itself in the vanguard of the

Figure 8.1 *Advertisement for a public procurement training course*

Green State. The key point to make here is that the power of purchase, like all powers, needs to be properly resourced before it can be effectively deployed. Among other things, this means enabling the public procurement profession to develop the competence and the confidence to help the public sector buy into sustainability.

Calibrating supply and demand

Creative procurement can play a genuinely innovative role in helping to calibrate supply and demand, a planning issue that seems to elude the Green State debate, despite its enormous significance. Where it has been most successful, school food reform has set new standards for the food service sector, creating new market opportunities for fresher, more nutritious produce, including organic produce, as the examples of Rome and East Ayrshire in particular show. However, unless these efforts are paralleled by supply-side measures to help local producers, especially small producers, a demand-side stimulus could invite a flood of imports, which is precisely what has happened with the development of the organic food market in the UK. Calibrating supply and demand is even more challenging in the developing country context. As we saw in Chapter 7, procurement agencies like the WFP are trying to use the power of purchase to fashion a supply-side capacity with small farmers who have hitherto been tethered to the world of subsistence agriculture. Calibrating supply and demand requires a set of skills, new commercial opportunities and prospects for producer collaboration that are seldom found inside the traditional public sector.

Disseminating good practice

The patchy nature of the school food revolution highlights a point that resonates more widely: good practice is a bad traveller (Morgan and Morley, 2006). If good practice is to become the norm rather than the exception, the public sector will need to devise more creative and more effective diffusion mechanisms. These will take different forms in different countries, including the establishment of professional associations and area-based associations to promote knowledge transfer between organizations on the one hand and localities on the other. Perhaps the key point to note about *social* learning is that the most effective diffusion mechanism appears to be horizontal peer-to-peer learning, rather than vertical command-and-control systems.

Governance structures

State governance structures pose both internal and external barriers to sustainable development. Internally, the multi-level governance structures of the state make it increasingly difficult to synchronize public policy between central and local levels, creating a disjuncture between design and delivery – a problem that needs to be addressed through more integrated and 'joined-up' policy processes. Externally, the structures of the state need to become more porous to facilitate genuine public participation in (and ownership of) public services. As well as helping to deepen democracy, which is an end in itself of course, greater public participation is also a means to making democracies work more effectively by enhancing their problem-solving capacities. If the 20th-century welfare state took human behaviour as a given, and sought to design policies *for* citizens, the 21st-century Green State will have to try to design policies *with* its citizens, in a shared endeavour to promote more sustainable modes of working and living.

To meet the most challenging issues of the 21st century – hunger, obesity and climate change, for example – the state can no longer pretend that it can solve them on its own. If it is to secure the active consent and problem-solving skills of civil society and the corporate sector, the state will need to fashion new, more iterative governance structures – both internally, to encourage more holistic policy processes, and externally, to harness the skills and ingenuity of society at large. As enlightened officials and politicians have come to realize, governments cannot meet the challenge of sustainable development alone.

Spatial fetishism

Perhaps the most important spatial lesson to emerge from the school food reform experience is the need to transcend the binary thinking that dominates so much of the debate about sustainable development. In extreme cases, this attaches certain attributes to particular spatial scales – for example, benign attributes to the 'local' scale and malign attributes to the 'global' scale. The idea of the 'local trap' we discussed in Chapter 1 helped to expose some of the worst examples of binary thinking in the sustainable food movement, where localization is extolled over globalization as a matter of principle. However, it is always worth repeating that 'globalization' and 'localization' are Janus-faced terms that need to be understood in the concrete, not the abstract, before we can assess their impact in sustainability terms. For example, globalization can be negative and regressive if it means that developing countries have to sign up to WTO rules (where they have less discretion over their domestic food security arrangements), because this would elevate neo-liberal trade principles over the social and economic needs of the poor. However, globalization can also be positive and progressive if it means a more concerted effort on the part of the international community to honour the MDGs.

Being clear-cut, these examples do not capture the more difficult decisions that have to be made by people who consider themselves to be 'ethical consumers'. One of the most difficult issues for such consumers is how to assess the carbon footprint of a product. The tendency is to use food miles as a guide, when in fact this is merely one component of the life cycle of a product. A decision based on food miles might persuade consumers to make a 'local' food choice, while a decision based on the life cycle could favour a 'global' food product. This is just one example of the tensions within the 'ethical foodscape', where consumers might feel torn between what they perceive to be 'local and green' purchases on the one hand and what they perceive to be 'global and fair' purchases on the other. In contrast to spatial fetishism, which makes dubious claims about places and products, these complex examples highlight the sheer cognitive limits to making decisions about what is more or less ethical, or more or less sustainable.

These are six dimensions from the school food reform experience that deserve to be given more prominence in the Green State debate, which often tends to sacrifice the particularity of the concrete for the generality of the abstract. After all, it is at the level of the concrete – in the seemingly prosaic decisions that are made in public services like school food – that the real tensions between quality and cost are played out. If the Green State debate is unable to engage with this concrete realm, to shed light on the present and to make hope practical for the future, then it will cut itself off from the warp and weft of everyday life and will foreclose the options for change for its communities.

Community food planning: Cities and the public plate

As one of the most used and abused terms in the political lexicon, 'community' always needs to be treated with due care and attention. New community schemes invariably attract questions about spatial scale: how is the community constituted, by whom and for what purpose? Like the debate about how local is local, these questions tend to generate fuzzy answers, because so much depends on context and purpose.

Because it underlines the deep interdependencies between peoples and places across the globe, the sustainable development paradigm has no place for the rigidly bounded spaces associated with the 'territorial trap'. The territorial trap refers to a set of unwarranted geographical assumptions that conjure up a world composed of nationally bounded territories, where the territorial state acts as geographical container of modern society. Critics argue that territorial scale is a socially constructed arrangement; as such, it is more fluid and more relational than is allowed for in the highly bounded conception (Agnew, 1994).

Although the assumptions of the territorial trap leave much to be desired from a sustainability perspective, which stresses interdependence rather than independence, it would be naïve to think that this bounded conception will soon disappear. On the contrary, the bounded view of the world has strong sponsorship from both ends of the political spectrum: some right-wing parties want to protect their domestic spaces from immigration, while some left-wing ecologists want to protect the 'carrying capacity' of their bioregions or promote the 'urban ecological security' of their cities – approaches that suggest the growth of a new territorial autarky.

A bounded view of the world can also be found in some quarters of the planning profession, which has signally failed to address one of the basic essentials of human life: the food system (Pothukuchi and Kaufman, 2000). To its credit, the American Planning Association (APA) is now addressing the bounded conception of space and the mysterious absence of food from the planning system:

> *Food is a sustaining and enduring necessity. Yet among the basic essentials for life – air, water, shelter and food – only food has been absent over the years as a focus of serious professional planning interest. This is a puzzling omission because, as a discipline, planning marks its distinctiveness by being comprehensive in scope and attentive to the temporal dimensions and spatial interconnections among important facets of community life.* (APA, 2007, p1)

The APA has moved a long way in a short time. At its national planning conference in San Francisco in 2005, for the first time in its history, the APA included food planning in a special track of sessions. By the summer of 2007, the APA had approved a bold and innovative policy guide 'for planners to become engaged in community and regional food planning' (APA, 2007). The following factors persuaded the profession that 'food planning' was an idea whose time had come:

- recognition that food system activities take up a significant amount of urban and regional land;
- awareness that planners can play a role to help reduce the rising incidence of hunger on the one hand and obesity on the other;
- understanding that the food system represents an important part of community and regional economies;
- awareness that the food people eat takes a considerable amount of fossil fuel energy to produce, process, transport and dispose of;
- understanding that farmland in metropolitan areas is being rapidly lost;
- understanding that pollution of ground and surface water, caused by the overuse of chemical fertilizers and pesticides in agriculture, adversely affects drinking water supplies;
- awareness that lack of healthy foods in low-income areas is an increasing problem that urban agriculture can address; and
- recognition that many benefits emerge from stronger community and regional food systems. (APA, 2007)

Having recognized the multidimensional significance of the food system, the APA's policy guide furnishes a new vision for the role of planning in a sustainable society. Especially significant is the fact that this vision tries to transcend the binary thinking that mars much of the work on local and community food policy, particularly as regards the global/local, the social/spatial and the conventional/alternative food sectors, as we discussed in Chapter 1. Let us briefly look at the APA's argument to see how it is trying to transcend these binary challenges.

Localization/globalization

To promote food systems that are ecologically sustainable, the APA subscribes to more food system localization. A core principle of sustainability, it says, is to meet human needs in ways that are 'as spatially proximate to their consumption as possible'. Communities that rely on distant food sources are rendered vulnerable to the vagaries of markets, transport and energy, over which they have little control. Greater self-reliance in food will help people to reconnect to their region and persuade them 'to care more about the region's resources'. This is a very strong endorsement of the localization principle to promote greater ecological self-reliance, based on the belief that spatial proximity fosters a stronger ethic of care.

To ensure that localism does not degenerate into parochialism, the APA takes a strong stand on international development. In an increasingly interdependent world, it says:

> *It is not only incumbent upon wealthier countries to act to end hunger and food insecurity across the globe, it is also important to redress the adverse impacts of agriculture trade policies on the ability of poor urban and rural households to subsist.*

To this end, the APA calls on the planning profession to support policies that promote development, rather than just aid.

Social/spatial

One of the standard criticisms of community-based planning is that it privileges the spatial over the social, downplaying the fact that class or ethnic groups may have very different needs despite sharing the same place. The APA is very conscious of this problem. Noting that 11 per cent of all US households were classed as 'food insecure' in 2005 (a proportion that rose to 17.9 per cent among Hispanic households and 22.4 per cent among Black households), the guide emphasizes that planners are 'uniquely positioned to help improve low-income people's access to programmes and facilities that enhance food security'. Through more imaginative and integrated policies for land use, transport and urban agriculture, for example, the APA believes that planners could play a much bigger role in addressing diet-related illnesses and food insecurity problems.

Conventional/alternative

As we argued in Chapter 1, there is a strong tendency for environmental groups and scholars to focus their efforts entirely on the alternative food sector, excluding the conventional food sector because it is deemed to be too industrialized. The APA avoids this binary thinking by focusing on both sectors of the food economy. Specifically, the guide offers support for planners 'to engage in planning that both strengthens community and regional food systems and encourages the industrial food system to provide multiple benefits to local areas'. This is a sound and sensible policy for planners to follow, because it helps to raise the quality of the mainstream food system, where the vast majority of consumers buy their food.[12]

In sum, the APA policy guide highlights the need to nurture a new form of planning – *community and regional food planning* – which can help to turn the power of the public plate from rhetoric into reality. Of all the players in community food planning, cities have the capacity to be the most influential, not least because of their large purchasing budgets and their growing significance as consumption centres. Indeed, as our case studies of Rome, London and New York indicate, some cities are becoming more proactive in this field because of the twin pressures of urban food security and sustainable development. In the EU, for instance, cities are waking up to the fact that their procurement budgets – their power of purchase – constitute a huge untapped resource. As much as 65 per cent of the public procurement market in the EU is now in the hands of sub-national public bodies as a result of devolution and decentralization of responsibilities to local and regional government (Eurocities, 2005). The C40 group of world cities – a new partnership pledged to reduce carbon emissions and increase energy efficiency in large cities across the world – has also made a priority of deploying its collective procurement power to promote sustainable development. To this end, eight priority sectors have been identified in which to exercise the power of purchase, including energy, transport, water and waste. Despite the fact that food is a major factor in each of these priority sectors, it does not figure in the thinking of the C40 alliance of world cities, leaving a vacuum at the heart of their sustainability plans.

Two very different planning scenarios can be identified for cities in the developed and developing worlds. In the developed world, cities will come under mounting pressure to demonstrate their ecological credentials – by aiming to become carbon neutral, for example. In the industrialized countries, some world cities are becoming ever more concerned about their 'urban ecological security' – that is, their capacity to secure the economic and ecological resources they need to become both more competitive and more sustainable. Cities, in this view, are endeavouring to become more autarkic – making less use of external resources and more use of 'enclosed' resources (Hodson and Marvin, 2007). If a new ecologically driven localism were to emerge in these world cities, we must hope it is not a defensive, self-referential form of localism – parochialism in disguise, in other words.

Cities in the developing world, by contrast, are facing a totally different kind of challenge: how to feed their burgeoning populations, particularly in the mega-cities that have already become mega-slums (Davies, 2006). The mega-cities of Mumbai, Dhaka, Lagos and Sao Paulo, for example, will soon have to feed more than 20 million people each, a task that could overwhelm them unless they receive more and better support from the developed world. Yet the FAO believes that the food dimension of urban poverty is not receiving the attention it deserves from global donors or local politicians and planners.

To address the growing problems of urban food insecurity and malnutrition and to build new partnerships, especially with local municipalities, the FAO launched its 'Food for the Cities' initiative in 2001. There is probably nowhere on Earth where community food planning is more urgently needed, or more difficult to implement, than in the teeming mega-cities of the South. Food security would be immeasurably improved if foreign aid was better targeted, if stronger urban–rural linkages were forged to enable cities to become engines of rural development and if the informal food sector, particularly the street vendor, was treated as a partner, rather than a pariah (FAO, 2000 and 2007b).

From the largest cities to the smallest counties, community food planning could be a vehicle to enhance the power of the public plate by extending the school food revolution to larger, more significant social and spatial scales. By deploying the power of purchase to wider communities of adults as well as children, the public plate could be harnessed by the state, particularly a Green State, to honour the most basic of all human rights: the right to food. If sufficient political will could be mustered for a new ethic of care, an ethic that had a global as well as a local reach, community food planning could really come of age, helping to deliver the *intrinsically* significant benefits of sustainable development.

Notes

Chapter 1

1 Agenda 21 is an action plan for sustainable development adopted by 178 governments at the 1992 United Nations Conference on Environment and Development (also known as the Earth Summit) held in Rio de Janeiro. Chapter 28 of Agenda 21 focuses on the role of local political authorities as facilitators in promoting sustainable development within their localities. Their contribution is emphasized particularly in relation to the specific responsibilities that local political authorities have in developing and maintaining local economic, social and environmental infrastructure, overseeing planning and regulations, implementing national environmental policies and regulations, and establishing local environmental policies and regulations (Baker, 2006, p106).

2 This model is based on a community of individuals that support a farm operation by covering its anticipated costs and the farmer's salary. In exchange, they receive shares of the farm's produce throughout the growing season. The fundamental idea behind Community Supported Agriculture is that growers and consumers share both the risks and the benefits of food production.

3 It must be stressed, however, that rescaling can also operate in the opposite way. For example, in the US large agri-business firms, faced with county opposition to concentrated animal feeding operations (CAFOs), have often sought to push the issue up to the state level, where their networks of influence are stronger. In many states, 'the approval process for large-scale livestock facilities has been taken from local authorities at the behest of major farm groups. Instead, the state agriculture departments decide, and they generally favour large-scale agriculture. In Illinois, for instance, the agriculture commissioner is a hog farmer who supports large-scale livestock operations' (Jones and Martin, 2006).

4 To his credit, Gareth Edwards-Jones has been making this argument for quite some time. He famously earned the wrath of green campaigners when he entered the BBC's Green Room to argue against food miles. 'While those making these calls may seem to have common sense on their side,' he said, 'the science which could be used to underpin their arguments is at best confusing, and at worst absent' (Edwards-Jones, 2006).

Chapter 2

1 Examples of the enormous potential of the public procurement market in Europe have been provided in a study conducted by the International Council for Local Initiatives between 2001 and 2003. According to this study, if all public authorities in the EU switched to green electricity, they would save more than 60 billion tonnes of CO_2; if they also used energy-efficient desktop computers, another 830,000 tonnes of CO_2 would be avoided (Day, 2005, p202).

2 The Corruption Perception Index has been compiled by Transparency

International, a global civil society organization, founded in 1993, which is involved in the fight against corruption around the world.

3 It must be emphasized, however, that, according to the results of a survey conducted amongst EU member states before the 2004 enlargement, only 19 per cent of public administrations practise a significant amount of *green* procurement. The highest percentages were in Sweden, Denmark and Germany (Day, 2005, p202).

4 Currently, the GPA has a total of 28 members, including Canada, the EU, Japan and the US.

5 The Internal Market and Services Directorate General is one of 37 Directorates General and specialized services of the European Commission. Its main role is to coordinate the Commission's policy on the Single Market, which aims to ensure the free movement of people, goods, services and capital within the European Union.

Chapter 3

1 In 1937, 15 states had passed legislation authorizing local school boards to operate lunchrooms and to serve meals at cost. Four of these states (Indiana, Vermont, Missouri and Wisconsin) made special provisions for needy children (Gunderson, 2007).

2 The Consumer and Marketing Service is a predecessor agency to the Food and Nutrition Service.

3 The consequences of this decision on American schoolchildren cannot be over-emphasized. A 2005 government survey showed that food sold outside the regulations of the NSLP is present in 83 per cent of elementary schools, 97 per cent of middle schools and 99 per cent of high schools (CSPI, 2004). Another survey, conducted in 2004, revealed that 75 per cent of beverage options and 85 per cent of snacks in school vending machines were of poor nutritional value (Simon, 2006, pp220–221).

4 Currently, for example, New York City has a citywide contract with Snapple, which allows only Snapple branded products in vending machines throughout the schools. For a thorough discussion of the consequences of pouring rights contracts and vending machines on children's health, see Nestle (2002).

5 In a system where pupils' participation is the single most important factor determining the amount of funding schools receive for their meals, the FNS emphasizes public information (through, for example, the development of programme materials and the organization of dissemination events) to assist states in their continuous efforts to expand the number of schools and children participating in the nutrition programmes.

6 To give an idea, since 2003 New York State has been allocated over US$3.1 million per year by DoD Fresh.

7 It is important to notice that the DoD Fresh has also been helpful in assisting the establishment of Farm-to-School programmes by creating relationships and networks between local producers and local food service personnel through pre-existing DoD Fresh networks (USDA Food and Nutrition Service, 2005).

8 To break even, many school districts in the US have turned to private food service management. New York City is one of the few that has decided to maintain self-operation – or, in other words, to keep school meals as an in-house service.

9 The city government is responsible for correctional institutions, libraries, public safety, recreational facilities, sanitation, water supply, welfare services and public

education. It is important to note that the city's public school system, which is managed by the New York City Department of Education and is responsible for as many as 1.1 million pupils, is the largest in the US.

10 Contributing to the deficit were also students who did not qualify for a free lunch or had not completed the paperwork but still gained free lunches by evoking sympathy from school lunch workers. According to the Department of Education, these freeloading students were costing them as much as US$5 million annually in federal reimbursement (Andreatta, 2006).

11 In New York, the Food Technology Department of SchoolFood decides which food products can be included in the school menus on the basis of their suitability and nutritional value and then writes a specification that describes the qualities of each product (including nutritional qualities, the cut or shape of individual ingredients, serving size and packaging requirement). The food is purchased through two main procurement mechanisms. For the most part, SchoolFood bids for private distributors, who must prove that they can purchase the specified items (at a price that will be agreed to in the contract) and that they have the warehousing, distribution and accounting capacity to fulfil the orders for all the schools in their districts. The second mechanism, called 'contract direct', allows SchoolFood to directly deal with manufacturers and negotiate a price with them.

12 Between November 2005 and October 2006, SchoolFood procured more than 1.1 million pounds of sliced apples, adding US$1,044,557 to the local agricultural economy (Market Ventures et al, 2007, p85).

Chapter 4

1 These include Friuli-Venezia Giulia, Veneto, Marche, Basilicata, Toscana and Emilia-Romagna.

2 Under this type of tendering system, the contractor sets up a price for a product or service. Bidders are required to offer a price for the proposed product or service that is lower than the price set up by the contractor. Contracts are awarded to the bidder that offers the lowest (in percentage) price.

3 As specified in the tender, contracted companies are expected to organize training courses for teachers and informational campaigns for children and parents around six main themes: 'quality' and the main features of the tender; food and lifestyle; the socio-psychological aspects of food; food and multiculturalism; children and food choices; and improving children's food habits (Comune di Roma, 2001).

4 On the basis of the European Geographical Indication system, introduced in 1993 (Council regulation (EEC) No. 2081/92), a Protected Denomination of Origin (PDO) product has been produced, processed and prepared within a certain geographical area that is exclusively linked to the quality and characteristics of the product. A Protected Geographical Indication (PGI) product has been produced, processed *or* prepared within a specific geographical area, which attributes to the product its main quality characteristics.

5 This is the internationally recognized standard for the quality management of businesses. It applies to the processes that create and control the products and services supplied by an organization.

6 'Hazard Analysis Critical Control Points' is a food safety methodology that relies upon the identification of 'critical control points' in food production and preparation processes.

7 In 2008, all school food in Rome was organic except fish and meat.

8 Rome purchases €6 million of Fair Trade products for its schools every year.
9 In Italy, the average price of a school meal is €4.30. Of this, €1.33 covers the costs of the food ingredients; €1.24 is spent on the distribution of the food; €0.82 includes industrial costs and the profit margin; €0.31 is the cost of personnel; €0.26 is spent on disposable plates; €0.25 covers the transport costs; and €0.04 is the cost of cleaning materials.

Chapter 5

1 This was a highly prescient analysis of the school meals question. In fact, unlike most critics of the reform, who merely wanted a restoration of the spending cuts, Lang argued that the quality and *culture* of school food were equally important. This campaign was one of the reasons why Tim Lang entered the arena of food (as distinct from agricultural) politics in the UK.
2 The 'Jamie Oliver effect' triggered the change in government policy, but the climate of opinion was already changing as a result of a whole series of factors, including the growing moral panic about obesity, the escalating costs of diet-related disease and a rash of food scares, all of which prepared the ground for the celebrity chef's message (Morgan, 2006).
3 As the transition is proving to be longer (and much more difficult) than originally thought, the Labour Government announced a new school lunch grant of £80 million a year for an additional three years up to 2010/2011. This was especially designed to cover the higher costs associated with fresh ingredients, labour time and new equipment.
4 Currently in the UK, free school meals are available to children attending publicly maintained schools if the parents receive one of the following benefits: Income Support; Income-based Jobseeker's Allowance; Child Tax Credit (with an annual income of £14,495 or less); and Support under Part 6 of the Immigration and Asylum Act 1999.
5 Hull became the first council area in the UK to offer free school meals when its Labour-controlled council introduced a three-year pilot scheme in 2004, which pushed the average take-up rate from 36 per cent to 65 per cent (Colquhoun 2007; Pike, 2007). However, a new Liberal-Democrat council abolished the scheme before the pilot had been evaluated, saying that the £3.8 million a year was not money well spent. The new council defended its decision by arguing that healthy school meals should not be confused with free school meals (Smith, 2006).
6 First introduced in 2003, the congestion charge triggered the most dramatic changes in commuting patterns of any large city in the world, inspiring interest and imitation all over the world. Just as important, though much less visible, is the Mayor's strategy of promoting walking, cycling and physical activity, all of which are part of a new drive to render London a healthier and less obesogenic urban environment (Mayor of London, 2008a).
7 This terrain was all mapped by the London Food Commission, a think-tank set up by the Greater London Council, which produced many reports in 1984–1990 stressing multicultural and social inclusiveness as goals for policies on nursery food, school meals catering education and food for ethnic minorities.
8 In Greenwich the school meal is worth £1.40 a day at nursery, primary and special schools and £1.30 at secondary schools.
9 It should be pointed out that Greenwich officials said that Oliver ran way over budget, a liberty afforded to a celebrity chef but denied to real dinner ladies.

10 Up until 1990, all schools in London were run by the Inner London Education Authority (ILEA).

Chapter 6

1 The Caroline Walker Trust was established in 1987 to improve public health through good food in the UK. Its first report, 'Nutritional guidelines for school meals', produced in 1992, was long considered the definitive document for nutrient-based standards for school food in the UK. The standards, which include values for energy, macronutrients and micronutrients, were updated in 2005 (see Crawley, 2005).

2 The story of the healthy fruit tuck service has striking similarities with New York City's experiment with sliced apples. After realizing that the schools were selling an average of just 30 apples a year, catering managers started selling *chopped* apples, which initially were topped with a little chocolate-flavoured cherry (gradually eliminated). In 2006, they sold 750,000 portions of apples through the fruit tuck service.

3 Carmarthenshire was benchmarked against its neighbouring local authorities in South Wales. Significantly, in 2005 these local authorities were hit by a serious *E. coli* outbreak, which spread across 44 schools, affecting 150 people and killing a five-year-old. The origins of the outbreak were traced back to a butcher's premises that had quite a long history of food safety and hygiene issues. The public inquiry that took place in 2008 revealed that the councils of Bridgend, Merthyr and Rhondda Cynon Taf had decided to overlook the concerns about the butcher's business practices raised by an environmental health officer in 2002 because, according to official documents, he made 'the lowest overall offer' (Brindley, 2008). The meaning and implications of this low-cost catering culture are examined in depth in the final chapter of the book.

4 Section 121 of the Government of Wales Act (1998) placed a legal requirement on the newly formed Welsh Assembly Government to promote sustainable development. It was the first time that a government in the EU was given such a duty.

5 In general, as we mentioned in the previous chapter, Wales lags behind Scotland and England in terms of school food reform. In fact, its reform strategy, *Appetite for Life*, was only launched in mid-2006 (four years after the Scottish strategy and three years after the English reform document) and commits a disappointing amount of £6.6 million to improving the school meal service between 2008 and 2010. Significantly, as we will discuss in Chapter 8, nutritional standards have not been transposed into legislation by the Welsh Assembly Government, which, in contrast, for example, with the approach of the Scottish Government, decided to *pilot* the introduction of the standards in four local authorities to assess their costs and implications. Carmarthenshire is not one of these local authorities.

6 In 1998, the Scotland Act devolved new legislative powers and governance arrangements to the newly created Scottish Parliament. Since then, devolution has effectively made the UK a multi-national state.

7 In 2005, Scotland was at the bottom or in the bottom two of a number of European league tables in terms of its premature death rate (which is 30 per cent higher than in England and Wales), incidence of low birth weight, number of infant deaths and number of under-age pregnancies. Obesity and overweight were also very high. For example, data published at the end of 2005 showed that 20 per cent of children aged three were overweight and 20 per cent of school-aged children were obese (see Lang et al, 2006, pp18–22).

8 On average, East Ayrshire has received £700,000 a year from *Hungry for Success*. Of this, £150,000 has been utilized by the Council to sponsor breakfast clubs that offer free breakfast to all school children.
9 Currently, 12 out of 15 products in East Ayrshire are sourced from within 40 miles and 50–60 per cent of the vegetables purchased are grown in Scotland.

Chapter 7

1 There are approximately 100 million primary school children not attending school, and two out of three of them are girls. Girls also form the majority of the 150 million primary school-age children who begin school but drop out before completing four years of education. This means that they have not acquired even basic literacy and numeracy skills. Of the 875 million illiterate adults in the world, two-thirds are women. We briefly discuss the implications of this situation in the final chapter of the book.
2 For example, although the UK and the European Commission are major donors, according to Save the Children (2007) their investments actually amount to less than 1 pence per day for every malnourished child from the former and 1.7 pence per day from the latter.
3 This emphasis on parents' involvement provides an interesting comparison with the Italian Canteen Commissions and with the School Food Partnership in New York. Of all the school meal systems discussed in this book, the British is then the only one that does not provide formal mechanisms to include parents.
4 In 2005, for example, the WFP collaborated with more than 2200 NGOs in 74 countries in almost 200 projects around the world.
5 Over the past 40 years, the WFP has handed over its SFPs to more than 30 countries, including Ecuador and Morocco in 2004 and China and the Dominican Republic in 2005. In some cases the WFP helps experienced countries to help less experienced ones: for example, it is currently helping to develop a partnership amongst Cape Verde, Angola, Mozambique and the Government of Brazil so that they can benefit from Brazil's long-standing expertise in school feeding. Another example, as we will see, is the way in which Ghana is used as an exemplar for other African countries.
6 From the outset, then, the stated intention of the legislation was to use food aid to promote trade and Japan proved to be the most successful example of this trade-through-aid philosophy. In 1954 Japan received nearly US$400 million worth of food aid, but just 21 years later it was buying over US$20 billion worth of US foodstuffs. Charitable SFPs were perceived to play a key role in creating new commercial markets for US products. As Senator George McGovern said in 1964, 'Japanese schoolchildren who learned to like American milk and bread in US-sponsored school lunch programmes have since helped to make Japan our best dollar purchaser for farm products' (George, 1986, p199).
7 The tide of opinion is clearly against the US system of in-kind food aid, and some major donors, like the EU, Canada and Australia, have been persuaded to convert some or all of their food aid programmes to cash donations. Indeed, among the major donors only the US, China and Japan have so far refused to convert to cash donations (Renton, 2007).
8 In its 2008 World Development Report – significantly called *Agriculture for Development* – the World Bank recognized that the agricultural sector had suffered from neglect and underinvestment for the past 20 years, bequeathing an astonish-

ing legacy: while 75 per cent of the world's poor live in rural areas, a mere 4 per cent of foreign aid goes to agriculture in developing countries. Echoing the *Halving Hunger* analysis, the World Bank concluded that the agricultural sector must be placed at the heart of the development agenda if the goals of halving extreme poverty and hunger by 2015 are to be realized (World Bank, 2006).

9 Zimbabwe is a classic case in point. Hitherto a land of plenty, this southern African state was tragically transformed from a breadbasket to a hunger hot spot in a very short time, largely because of a catastrophic failure of state governance under the regime of Robert Mugabe.

Chapter 8

1 It is no coincidence that Finland and Sweden have the highest take-up rates in the Western world – at 90 per cent and 85 per cent respectively (School Food Trust, 2006).

2 Cooper says that she answers to three masters: the school district is her official employer, the USDA subsidizes her meals and the Chez Panisse Foundation, set up by chef Alice Waters, pays her salary of US$95,000 a year plus benefits. The first needs her to stay within budget, the second requires her to meet its dietary standards and the third wants her 'to hurry up and start a revolution' (Bilger, 2006).

3 As we mentioned in Chapter 6, the Welsh Assembly Government decided that Wales would not introduce the nutrient-based food standards that had been adopted in England and Scotland because school caterers argued that, if introduced too quickly and without consultation, they would reduce take-up rates even further, jeopardizing the viability of the service. In England, local authorities have had to report their take-up rates more rigorously since a new national indicator (NI 52) was introduced in 2007 (Baines, 2008).

4 Having a local supplier does not necessarily mean using local food, as the case of Carmarthenshire shows. By the same token, a national vendor can become a supplier of local food. As always, it is a case of looking beyond the label.

5 Personal communication from Marion Kalb, who also said that kitchen equipment in every school would provide a big nationwide boost to the Farm-to-School movement.

6 The following data are drawn from the WHO website: www.who.int/dietphysicalactivity/childhood.

7 The citizens' panel was convened by the Chronic Disease Prevention Alliance of Canada (CDPAC) as part of its Policy Consensus Forum on Obesity and the Impact of Marketing on Children.

8 The American food industry has also undermined the WHO's anti-obesity campaign. One of the most notable examples of its global reach and influence came in 2003, when it challenged a draft WHO report called *Diet, Nutrition and the Prevention of Chronic Diseases*, which recommended that no more than 10 per cent of a person's total calorie intake should come from 'added sugars'. Far from simply contesting the science, the Sugar Association warned that it would try to persuade the US Government to cut its contributions to the WHO unless the draft report tempered its message on sugar (Alden and Buckley, 2004).

9 To put these sums in perspective, we need to remember what the US Government is spending on the Iraq war. The most conservative estimate (from the Congressional Budget Office) said that US$413 billion had been spent by 30 September 2007, a figure that could increase to US$2000 billion by 2017 (Fidler,

2008). This means that the US Government would need to devote just one per cent of what it is spending on the war annually to feed schoolchildren in Africa.

10 Unless otherwise stated, we draw on this ActionAid report here.

11 Women's lack of access to land, for example, is one of the key obstacles to tackling hunger and meeting the first MDG. Women produce up to 80 per cent of food in developing countries, yet female-headed households are more likely to suffer chronic hunger. For these women, land rights are crucial to securing their right to food.

12 It is impossible to ignore the conventional sector, not least because, the guide says, the top five food retailers accounted for some 43 per cent of sales in 2000, up from 24 per cent in 1997.

References

ActionAid (2008) *Hit or Miss? Women's Rights and the Millennium Development Goals*, Action Aid, London

Agnew, J. (1994) 'The territorial trap: The geographical assumptions of international relations theory', *Review of International Political Economy*, vol 1, no 1, pp53–80

Ahmed, A., Hill, R., Smith, L., Wiesman, D. and Frankenberger, T. (2007) *The World's Most Deprived: Characteristics and Causes of Extreme Poverty and Hunger*, IFPRI, Washington, DC

Alden, E. and Buckley, N. (2004) 'Sweet deals: Big sugar fights threats from free trade and a global drive to limit consumption', *Financial Times*, 27 February

Allaire, G. (2004) 'Quality in economics: A cognitive perspective', in M. Harvey, M. McMeekin and A. Warde (eds) *Qualities of Food*, Manchester University Press, Manchester, UK

Allen, P. and Guthman, J. (2006) 'From "old school" to "Farm-to School": Neo-liberalization from the ground up', *Agriculture and Human Values*, vol 23, no 4, pp401–415

Allen, P., FitzSimmons, M., Goodman, D. and Warner, K. (2003) 'Shifting plates in the agrifood landscape: The tectonics of alternative agrifood initiatives in California', *Journal of Rural Studies*, vol 19, pp61–75

Andreatta, D. (2006) 'School lunch crunch', *The New York Post*, 19 October

Anthony Collins Solicitors (2006) *The Scope for Using Social Clauses in UK Public Procurement to Benefit the UK Manufacturing Sector: A Report for the Manufacturing Forum*, Department of Trade and Industry, London

APA (2007) *Policy Guide on Community and Regional Food Planning*, American Planning Association, www.planning.org/policyguides/pdf/food.pdf

Audit Commission (2001) *Carmarthenshire County Council – Catering Service*, Best Value Inspection Service, Audit Commission, Cardiff, UK

Baines, M. (2008) 'The new national indicator set for local government: NI 52 – Take-up of school lunches', Briefing 08/01, Association for Public Service Excellence, Manchester, UK

Baines, M. and Bedwell, J. (2008) 'School meals trend analysis', Briefing 08/02, Association for Public Service Excellence, Manchester, UK

Baker, S. (2006) *Sustainable Development*, Routledge, London

Barham, E. (2003) 'Translating terroir: The global challenge of French AOC labelling', *Journal of Rural Studies*, vol 19, pp127–138

Barry, J. and Eckersley, R. (2005a) 'An introduction to reinstating the state', in J. Barry and R. Eckersley (eds) *The State and the Global Ecological Crisis*, MIT Press, Cambridge, MA, ppix–xxv

Barry, J. and Eckersley, R. (2005b) 'W(h)ither the Green State?', in J. Barry and R. Eckersley (eds) *The State and the Global Ecological Crisis*, MIT Press, Cambridge, MA, pp255–271

Beattie, A. (2007a) 'Food safety clash gives taste of battles ahead', *Financial Times*, 1 August

Beattie, A. (2007b) 'US Farm Bill Reform: Pile-it-high advocates set to reap gains',

Financial Times, 9 October
Beattie, A. (2008) 'Boom challenge for food aid policy', *Financial Times*, 7 February
Bellows, A. and Hamm, M. W. (2001) 'Local autonomy and sustainable development: Testing import substitution in localizing food systems', *Agriculture and Human Values*, vol 18, pp271–284
Bennett, J. (2003) *Review of School Feeding Projects*, Department for International Development, London
Bertino, R. (2006) 'Scuole e aziende, avanti con bio' ['Schools and firms, organics moves on'], *Ristorazione Collettiva*, April, pp36–41
Bilger, B. (2006) 'The lunchroom rebellion', *The New Yorker*, 4 September
Birchall, J. (2007) 'Foodmakers tighten code to avoid ads ban', *Financial Times*, 18 July
Black, J. (2007) 'Senate drops measure to greatly reduce sugar and fat in food at schools', *Washington Post*, 15 December
Born, B. and Purcell, M. (2006) 'Avoiding the local trap: Scale and food systems in planning research', *Journal of Planning Education and Research*, vol 26, pp195–207
Bouwer, M. et al (2006) *Green Public Procurement in Europe*, Milieu and Management, Virage, The Netherlands
Bowden, C., Holmes, M. and Mackenzie, H. (2006) 'Evaluation of a pilot scheme to encourage local suppliers to supply food to schools', Environment and Rural Affairs Division, Scottish Executive, Edinburgh
Brera, P. G. (2007) 'Distributori di frutta nelle scuole' ['Fruit vending machines in the schools'], *La Repubblica*, 2 March
Brescianini, S., Gargiulo, L. and Gianicolo, E. (2002) 'Eccesso di peso nell'infanzia e nell'adolescenza'['Overweight during infancy and adolescence'], paper presented at the ISTAT conference, ISTAT (National Institute of Statistics), Rome
Brescoll, V. L., Kersh, R. and Brownell, K. D. (2008) 'Assessing the feasibility and impact of Federal childhood obesity policies', *The Annals of the American Academy of Political and Social Science*, vol 615, no 1, pp178–194
Brindley, M. (2008) 'Councils chose "lowest cost" Tudor's meat despite scores of complaints', *Western Mail*, 13 February
Calhoun, C. (ed) (1992) *Habermas and the Public Sphere*, MIT Press, Cambridge, MA
Campbell, M. C. (2004) 'Building a common table: The role for planning in community food systems', *Journal of Planning Education and Research*, vol 23, pp341–355
Carmarthenshire Catering Services (2004) *Local Sustainable Food Strategy: Building Sustainable Development into Food Contracts*, Carmarthenshire Catering Services, Carmarthen, UK
Carmarthenshire County Council (2004) *Carmarthenshire Community Strategy*, Carmarthenshire County Council, Carmarthen, UK
Carmarthenshire County Council (2005) *Primary Menu*, Carmarthenshire County Council, Carmarthen, UK
Carmarthenshire Partnership (2004) *Carmarthenshire School Meals Nutrition Strategy: Improving the Health of Children and Young People in Carmarthenshire*, Carmarthenshire Partnership, Carmarthen, UK
Carter, N. (2007) *The Politics of the Environment: Ideas, Activism, Policy* (second edition), Cambridge University Press, Cambridge, UK
Cavallaro, V. and Dansero, E. (1998) 'Sustainable development: Global or local?', *GeoJournal*, vol 45, no 1, pp33–40
Cawson, A., Morgan, K. J., Webber, D. and Holmes, P. (1990) *Hostile Brothers: Competition and Closure in the European Electronics Industry*, Clarendon Press, Oxford, UK
CDPAC (2008) 'Panel calls for leadership to protect children from targeted marketing',

Chronic Disease Prevention Alliance of Canada, Ottawa
Ceccarelli, F. (2005) 'Veltroni e il modello Roma-comunitá' ['Veltroni and the Rome-community model'], *La Repubblica*, 23 November
Ceccarelli, L. (2006) 'Gusto, fantasia e solidarietà: Come mangiano gli Italiani' ['Taste, fantasy and solidarity: How Italians eat'], *La Repubblica*, 30 October
CFC (Children's Food Campaign) (2008) 'Children's food campaigners say "Well done on cooking lessons, now for junk food ads"', Sustain, London
CFSC (Community Food Security Coalition) (2007a) *Community Food Security News*, spring 2007
CFSC (2007b) 'Geographic preferences for schools: Connecting kids and communities', www.foodsecurity.org/GeogPreferencing-CFP1-pager.pdf
Chasek, P. S., Downie, D. L. and Welsh Brown, J. (2006) *Global Environmental Politics*, Fourth Edition, Westview Press, Cambridge, MA
Clement, C. (1996) *Care, Autonomy and Justice: Feminism and the Ethic of Care*, Westview Press, Oxford, UK
Cohen, N. (2007) 'The local trap', electronic mailing list, foodplanning@u.washington.edu
Collier, P. (2007) *The Bottom Billion: Why the Poorest Countries are Failing and What Can Be Done About It*, Oxford University Press, Oxford, UK
Collins, B. (2008) 'Farm-to-School for all', www.chefann.com/blog/?p=983
Colquhoun, D. (2007) 'The Hull experience in providing free healthy meals for all primary school children', paper presented at the Second Annual Conference, Free Healthy School Meals: The Hull Experience, Hull, UK, 17 November
Comune di Roma (2001) *Capitolato Speciale per la Gestione del Servizio di Ristorazione Scolastica del Comune di Roma, 2002–2004* [*Special Tender for the Management of the School Meal Service in the City of Rome, 2002–2004*], Dipartimento XI, Rome
Comune di Roma (2004a) *Capitolato Speciale per la Gestione del Servizio di Ristorazione Scolastica del Comune di Roma, 2004–2007* [*Special Tender for the Management of the School Meal Service in the City of Rome, 2004–2007*], Dipartimento XI, Rome
Comune di Roma (2004b) *A Scuola con Più Gusto: Il Servizio di Ristorazione nelle Scuole Romane* [*At School with More Taste: The School Meal Service in the Roman Schools*], Assessorato e Dipartimento XI Politiche Educative e Scolastiche, Rome
Comune di Roma (2007) *Capitolato Speciale per la Gestione del Servizio di Ristorazione Scolastica del Comune di Roma, 2007–2012* [*Special Tender for the Management of the School Meal Service in the City of Rome, 2007–2012*], Dipartimento XI, Rome
Cooke, P. and Morgan, K. J. (1998) *The Associational Economy: Firms, Regions and Innovation*, Oxford University Press, Oxford, UK
Cooper, A. and Holmes, A. (2006) *Lunch Lessons: Changing the Way we Feed our Children*, HarperCollins, New York
Craig, D. (2006) *Plundering the Public Sector*, Constable, London
Crawley, H. (2005) *Nutrient-Based Standards for School Food: A Summary of the Standards and Recommendations of the Caroline Walker Trust and the National Heart Forum*, The Caroline Walker Trust, Abbots Langley, UK
CSPI (Center for Science in the Public Interest) (2003) 'Pestering parents: How food companies market obesity to children', Center for Science in the Public Interest, Washington, DC
CSPI (2004) 'Dispensing junk: How school vending undermines efforts to feed children well', Center for Science in the Public Interest, Washington, DC
CSPI (2006) 'Raw deal: School beverage contracts less lucrative than they seem', Center for Science in the Public Interest, Washington, DC
CSPI (2007) 'Sweet deals: School fundraising can be healthy and profitable', Center for Science in the Public Interest, Washington, DC

Culinary Institute of America (2005) 'Diet and health: Hot issues for the foodservice industry', *Mise en Place*, vol 33, pp11–13

Cullen, E. (2007) 'Local sourcing for school meals: Carmarthenshire County Council', in V. Wheelock (ed) *Healthy Eating in Schools: A Handbook of Practical Case Studies*, Verner Wheelock Associates, Skipton, UK, pp221–225

Curtis, F. (2003) 'Eco-localism and sustainability', *Ecological Economics*, vol 46, pp83–102

Daly, H. E. (1996) *Beyond Growth: The Economics of Sustainable Development*, Bacon Press, Boston, MA

Davies, M. (2006) *Planet of Slums*, Verso, London

Day, C. (2005) 'Buying green: The crucial role of public authorities', *Local Environment*, vol 10, no 2, pp201–209

Defra (2003) 'Unlocking opportunities: Lifting the lid on public sector food procurement', Department for the Environment, Food and Rural Affairs, UK Government, London

DeLind, L. B. (2006) 'Of bodies, places and culture: Re-situating local food', *Journal of Agricultural and Environmental Ethics*, vol 19, pp121–146

Department of Health (2008) 'Government announces first steps in strategy to help people maintain healthy weight and live healthier lives', Department of Health, UK Government, London

Department for Education and Skills (2006) 'Setting the standard for school food', press release, Department for Education and Skills, UK Government, London, 19 May

Diamanti, I. (2006) 'La nuova era della sazietá' ['The new era of satiety'], *La Repubblica*, 30 October

Dobson, A. (2003) *Citizenship and the Environment*, Oxford University Press, Oxford, UK

Druce, C. (2007) 'Scolarest leads a withdrawal from school contracts', *Caterersearch*, 14 June

Drummond, I. and Marsden, T. (1999) *The Condition of Sustainability*, Routledge, London

DuPuis, E. M. and Goodman, D. (2005) 'Should we go "home" to eat? Toward a reflexive politics of localism', *Journal of Rural Studies*, vol 21, pp359–371

DuPuis, E. M., Goodman, D. and Harrison, J. (2006) 'Just values or just value? Remaking the local in agro-food studies', in T. K. Marsden and J. Murdoch (eds) *Between the Local and the Global: Confronting Complexity in the Contemporary Food Sector*, Elsevier, Amsterdam

Dwyer, J. (1995) 'The School Nutrition Dietary Assessment Study', *American Journal of Clinical Nutrition*, vol 61 (supplement), pp173–177

Dykshorn, A. (2007) 'The school meal system in New York City: An overview of resources on the Federal, State and City levels', unpublished report

Earth Council (1994) *The Earth Summit – Eco 92: Different Visions*, Earth Council and the Inter-American Institute for Cooperation on Agriculture, San Jose

East Ayrshire Council (2005) Procurement Section tender documents, East Ayrshire Council, Kilmarnock, UK

East Ayrshire Procurement Section (2005) *Specification. Supply and Delivery of Fresh/Organic Food Stuffs to 11 East Ayrshire Schools*, East Ayrshire Procurement Section, East Ayrshire Council, Kilmarnock, UK

Easterly, W. (2006) *The White Man's Burden: Why the West's Efforts to Aid the Rest Have Done So Much Ill and So Little Good*, Penguin Books, Harmondsworth, UK

Eckersley, R. (2004) *The Green State: Rethinking Democracy and Sovereignty*, MIT

Press, Cambridge, MA

Eckersley, R. (2005) 'Greening the nation-state: From exclusive to inclusive sovereignty', in J. Barry and R. Eckersley (eds) *The State and the Global Ecological Crisis*, MIT Press, Cambridge, MA

Edwards-Jones, G. (2006) 'Food miles don't go the distance', http://newsbbc.co.uk/1/hi/sci/tech/4807026.stm

Elliott, S. (2007) 'Straight A's, with a burger as a prize', *New York Times*, 6 December

Escobar, A. (2001) 'Culture sits in places: Reflections on globalism and subaltern strategies of localization', *Political Geography*, vol 20, pp139–174

Eurocities (2005) *The CARPE Guide to Responsible Procurement*, Eurocities, Brussels

European Commission (2004) Directive 2004/18/EC of the European Parliament and of the Council of 31 March 2004 on the Coordination of Procedures for the Award of Public Works Contracts, Public Supply Contracts and Public Service Contracts, *Official Journal of the European Union*, 30 April

FAO (2000) *Food for the Cities: Food Supply and Distribution Policies to Reduce Urban Food Insecurity*, Food and Agriculture Organization, Rome

FAO (2006) *The State of Food Insecurity in the World, 2006*, Food and Agriculture Organization, Rome

FAO (2007a) *The Right to Food*, Food and Agriculture Organization, Rome

FAO (2007b) *Promises and Challenges of the Informal Food Sector in Developing Countries*, Food and Agriculture Organization, Rome

Feagan, R. (2007) 'The place of food: Mapping out the "local" in local food systems', *Progress in Human Geography*, vol 31, no 1, pp23–42

Feenstra, G. (1997) 'Local food systems and sustainable communities', *American Journal of Alternative Agriculture*, vol 12, pp28–36

FFPP (Farm and Food Policy Project) (2007) *Seeking Balance in US Farm and Food Policy*, Farm and Food Policy Project, Washington, DC

Fidler, S. (2008) 'War's spiralling cost inspires shock and awe', *Financial Times*, 18 March

Finocchiaro, B. R. (2001) 'Il progetto interregionale: Un programma a tutto campo di comunicazione ed educazione alimentare' ['The inter-regional project: A wide range programme for dietary communication and education'], in B. R. Finocchiaro (ed) *La Ristorazione Scolastica: Prospettive Future* [*School Meals: Perspectives for the Future*], Quaderno 5, Cultura che Nutre, Programma Interregionale di Comunicazione ed Educazione Alimentare, Ministero delle Politiche Agricole e Forestali, Rome

Fisher, B. and Tronto, J. (1991) 'Toward a feminist theory of care', in E. Abel and M. Nelson (eds) *Circles of Care: Work and Identity in Women's Lives*, State University of New York Press, Albany, NY

Fisher, E. (2007) 'A desk review of the Ghana school feeding programme', in K. J. Morgan et al (2007) *Home Grown: The New Era of School Feeding*, World Food Programme, Rome

Food and Research Action Center (2007) 'Local school wellness policies', www.frac.org/html/federal_food_programs/programs/school_wellness.html

FoodManagement (2006a) 'Big changes in the Big Apple', *Food Management*, November, www.food-management.com/fm_innovator/fm_imp_15469

FoodManagement (2006b) 'Menu challenge: Changes in attitude, changes in latitudes', *Food Management*, November, www.food-management.com/fm_innovator/fm_imp_15466

Foresight (2007) *Tackling Obesities: Future Choices*, Government Office for Science, London

GAO (US Government Accountability Office) (2007) 'Various challenges impede the efficiency and effectiveness of US food aid', report to the Committee on Agriculture, Nutrition and Forestry, US Senate, Washington, DC

Gapper, J. (2007) 'NyLon, a tale of twin city-states', *Financial Times*, 25 October

Garland, S. (2006) 'Glass half empty at schools, milk choice advocates say', *New York Sun*, 13 December

Garnett, S. (2007) 'School Districts and Federal Procurement Regulations', United States Department of Agriculture, Alexandria, VA

General Services Administration, Department of Defense and National Aeronautics and Space Administration (2005) *Federal Acquisition Regulation*, vol 1, parts 1–51, www.arnet.gov/far/current/pdf/FAR.pdf

George, S. (1986) *How the Other Half Dies: The Real Reasons for World Hunger*, Penguin Books, Harmondsworth, UK

George, S. (1990) *Ill Fares The Land: Essays on Food, Hunger and Power*, Penguin Books, London

Gershon, P. (1999) *Review of Civil Procurement in Central Government*, HM Treasury, London

Gershon, P. (2001) Speech to 'Greening Government Procurement Conference', 22 May, London

Gibbon, D. and Jakobsson, K. M. (1999) 'Towards sustainable agricultural systems', in A. K. Dragun and C. Tisdell (eds) *Sustainable Agriculture and Environment: Globalization and the Impact of Trade Liberalisation*, Edward Elgar, Cheltenham, UK

Godwin, J. et al (2007) 'South Gloucestershire Economic Development Strategy 2007–2015' (consultation draft), December, South Gloucestershire Council, Community Services, Yate, UK

Goodman, D. (2004) 'Rural Europe redux? Reflections on alternative agro-food networks and paradigm change', *Sociologia Ruralis*, vol 44, no 1, pp3–16

Gottlieb, R. (2001) *Environmentalism Unbound: Exploring New Pathways for Change*, MIT Press, Cambridge, MA

Gourlay, R. (2007) 'Sustainable school meals: Local and organic produce', in V. Wheelock (ed) *Healthy Eating in Schools: A Handbook of Practical Case Studies*, Verner Wheelock Associates, Skipton, UK

Green Planet.Net (2006) 'Mense bio, un premio Europeo ai genitori di Budoia (PN)' ['Organic canteens: A European prize for Budoia's parents'], 26 January, www.greenplanet.net/content/view/13464

Greenwich Council (2007) *Healthy Communities Strategy 2006–2008*, Greenwich Council, London

Griffiths, J. (2006) 'Mini-symposium: Health and environmental sustainability. The convergence of public health and sustainable development', *Public Health*, vol 120, pp581–584

GSFP (Ghana School Feeding Programme) (2006) *The Fight Against Hunger*, newsletter, no 2, October

Gunderson, G. W. (2007) 'The national school lunch program: Background and development', www.fns.usda/gov/cnd/Lunch/AboutLunch/NSLP-Program%20History.pdf

Gustafsson, U. (2002) 'School meals policy: The problem with governing children', *Social Policy and Administration*, vol 36, no 6, pp685–697

Guthman, J. (2004) 'The trouble with "organic lite" in California: A rejoinder to the "conventionalization" debate', *Sociologia Ruralis*, vol 44, pp301–316

Guthman, J. and DuPuis, M. (2006) 'Embodying neoliberalism: Economy, culture and the politics of fat', *Environment and Planning D*, vol 24, pp427–448

Guy, C., Clarke, G. and Eyre, H. (2004) 'Food retail change and the growth of food deserts: A case study of Cardiff', *International Journal of Retail and Distribution Management*, vol 32, no 2, pp72–88

Habermas, J. (1989) *Structural Transformation of the Public Sphere*, MIT Press, Cambridge, MA

Hajer, M. (1995) *The Politics of Environmental Discourse*, Oxford University Press, Oxford, UK

Hamilton, N. (2002) 'Putting a face on our food: How state and local food policies can promote the new agriculture, *Drake Journal of Agricultural Law*, vol 7, pp407–424

Hamlin, A. (2006) 'School foods 101', *Vegetarian Journal's Foodservice Update*, vol XII, pp10–12

Hansard (1979) Parliamentary Debate on the Education (No 2) Bill, 5 November

Harrell, E. (2007) 'CARE turns down US food aid', *Time*, 15 August

Harvey, M., McMeekin, M. and Warde, A. (2004) 'Introduction: Food and quality', in M. Harvey, M. McMeekin and A. Warde (eds) *Qualities of Food*, Manchester University Press, Manchester, UK

Harvey, M., McMeekin, M. and Warde, A. (2004) 'Conclusion: Quality and processes of qualification', in M. Harvey, M. McMeekin and A. Warde (eds) *Qualities of Food*, Manchester University Press, Manchester, UK

Hatanaka, M., Bain, C. and Busch, L. (2006) 'Differentiated standardization, standardized differentiation: The complexity of the global agri-food system', in T. Marsden and J. Murdoch (eds) *Between the Local and the Global: Confronting Complexity in the Contemporary Agri-Food Sector*, Research in Rural Sociology and Development, vol 12, Elsevier, Amsterdam

Hay, C. (1996) 'From crisis to catastrophe? The ecological pathologies of the liberal-democratic state', *Innovations*, vol 9, no 4, pp421–434

Held, V. (2005) *The Ethics of Care: Personal, Political and Global*, University of Oxford Press, Oxford, UK

Helstosky, C. (2006) *Garlic and Oil: Food and Politics in Italy*, Berg, Oxford, UK

Hicks, K. M. (1996) *Food Security and School Feeding Programs*, Catholic Relief Services, Baltimore, MD

Hines, C. (2000) *Localization: A Global Manifesto*, Earthscan, London

Hinrichs, C. C. (2000) 'Embeddedness and local food systems: Notes on two types of direct agricultural markets', *Journal of Rural Studies*, vol 16, pp295–303

Hinrichs, C. C. (2003) 'The practice and politics of food system localization', *Journal of Rural Studies*, vol 19, pp33–45

HIPL (Harrison Institute for Public Law) (2007) *Helping Schools Buy Local: An Overview of the Issues*, Georgetown University, Washington, DC

HM Treasury (2007) *Transforming Government Procurement*, HM Treasury, London

Hodson, M. and Marvin, S. (2007) 'Urban ecological security: The new urban paradigm?', *Town and Country Planning*, December, pp436–438

House of Commons Health Committee (2004) 'Obesity', third report of Session 2003–04, the House of Commons, London

Ilbery, B. and Kneafsey, M. (2000) 'Producer constructions of quality in regional speciality food production: A case study from South West England', *Journal of Rural Studies*, vol 16, pp217–230

Imhoff, D. (2007) *Foodfight: The Citizen's Guide to a Food and Farm Bill*, University of California Press, Berkeley, CA

International Obesity Task Force (2005) 'EU Platform on diet, physical activity and health', EU Platform briefing paper prepared in collaboration with the European Association for the Study of Obesity, Brussels, 15 March

Jacobs, M. (1999) 'Sustainable development as a contested concept', in A. Dobson (ed) *Fairness and Futurity: Essays on Environmental Sustainability and Social Justice*, Oxford University Press, Oxford

Jones, T. and Martin, A. (2006) 'Hog wars: Missourians raise stink over giant operations', *Chicago Tribune*, 12 March

Joshi, A., Kalb, M. and Beery, M. (2006) *Going Local: Paths to Successes for Farm to School Programs*, developed by the National Farm-to-School Program Center for Food and Justice, Occidental College and the Community Food Security Coalition, http://departments.oxy.edu/uepi/cfj/publications/goinglocal.pdf

Kanemasu, Y. (2007) 'The New York City school meal system', unpublished report, School of City and Regional Planning, Cardiff University, Cardiff, UK

Karp Resources (2007) 'Schools globally are thinking locally', *News from the Strategy Kitchen*, summer

Kaufman, J. (2005) 'The role of planners in the emerging field of community food system planning', Louis B. Wetmore Lecture on Planning Practice, University of Illinois Planning Institute, Urbana-Champagne, IL, 2 March

Kaufman, L. and Karpati, A. (2007) 'Understanding the sociocultural roots of childhood obesity: Food practices among Latino families of Bushwick, Brooklyn', *Social Science and Medicine*, vol 64, pp2177–2188

Kington, T. (2007) 'Italians pay price for junk food revolution', *The Guardian*, 20 February

Kirwan, J. (2006) 'The interpersonal world of direct marketing: Examining conventions of quality at UK farmers' markets', *Journal of Rural Studies*, vol 22, pp301–312

Kloppenburg, J., Hendrickson, J. and Stevenson, G. W. (1996) 'Coming into the foodshed', *Agriculture and Human Values*, vol 13, pp33–42

Kloppenburg, J., Lezberg, J., DeMaster, K., Stevenson, G. W. and Hendrickson, J. (2000) 'Tasting food, tasting sustainability: Defining the attributes of an alternative food system with competent, ordinary people', *Human Organization*, vol 59, no 2, pp177–186

Knight, K. (2004) 'Increasing take-up in South Gloucestershire', in C. Hurley and A. Rile (eds) *Recipe for Change: A Good Practice Guide to School Meals*, Child Poverty Action Group (CPAG), London

Kwame, B. (ed) (2007) *Ghana: A Decade of the Liberal State*, Zed Books, London

Lang, T. (1981) *Now You See Them... Now You Don't: A Report on the Fate of School Meals and the Loss of 300,000 Jobs*, The Lancashire School Meals Campaign, Accrington, UK

Lang, T. and Heasman, M. (2004) *Food Wars: The Global Battle for Minds, Mouths and Markets*, Earthscan, London

Lang, T., Dowler, E. and Hunter, D. J. (2006) *Review of the Scottish Diet Action Plan: Progress and Impacts 1996–2005*, Health Scotland, Edinburgh

LDA (London Development Agency) (2006) *Healthy and Sustainable Food for London: The Mayor's Food Strategy Summary*, May, London, www.london.gov.uk/mayor/health/food/docs/food-strategy-summary.pdf

Leith, P. (2007) Speech to the LACA Annual Conference, Birmingham, UK, 12 July

Livingstone, K. (2006) Speech to 'Feeding our Cities in the Twenty-First Century', Soil Association Conference, The Brewery Conference Centre, London, 6–7 January

Lundqvist, L. J. (2001) 'A green fist in a velvet glove: The ecological state and sustainable development', *Environmental Values*, vol 10, pp455–472

MacMillan, T., Dowler, E. and Archard, D. (2004) 'Corporate responsibility for children's diets', in J. Gunning and S. Holm (eds) *Ethics, Law and Society*, vol 2, Ashgate, Aldershot, UK, pp237–243

Maisto, T. (2007) 'A scuola pasti etnici ed equosolidali' ['Ethnic and fair trade meals at school'], *La Repubblica*, 15 March

Mansfield, B. (2003) 'Spatializing globalization: A "geography of quality" in the seafood industry', *Economic Geography*, vol 79, no 1, pp1–16

Market Ventures, Inc., Karp Resources and New York University Center for Health and Public Service Research (2005) 'SchoolFood Plus evaluation: Interim report: Phase 2', Portland Market Ventures, Portland, ME

Market Ventures, Inc., Karp Resources and New York University Center for Health and Public Service Research (2007) 'SchoolFood Plus evaluation: Interim report: Phase 3, School Year 2005–2006', Portland Market Ventures, Portland, ME

Marquand, D. (2004) *Decline of the Public: The Hollowing-Out of Citizenship*, Polity Press, Cambridge, UK

Marsden, T. (2004) 'Theorizing food quality: Some key issues in understanding its competitive production and regulation', in M. Harvey, M. McMeekin and A. Warde (eds) *Qualities of Food*, Manchester University Press, Manchester, UK

Massimiani, L. (2006) 'Le mense romane: Un percorso di qualitá' ['The Roman school canteens: A quality process'], *InComune*, vol 11, nos 123–124, August–September, pp13–18

Mayor of London (2006) *The London Plan*, the Mayor's Office, London

Mayor of London (2008a) 'Mayor to lead on obesity and physical activity in London', press release, the Mayor's Office, London, 4 February

Mayor of London (2008b) 'London's regional government sets benchmark for fair procurement', press release, the Mayor's Office, London, 11 February

McGovern, G. and Quinn, C. (2006) 'Breakfast: The first meal toward ending hunger', *Metro*, June

McMichael, P. (2000) 'The power of food', *Agriculture and Human Values*, vol 17, pp21–33

Meadowcroft, J. (2007) 'Who is in charge here? Governance for sustainable development in a complex world', *Journal of Environmental Policy and Planning*, vol 9, no 4, pp299–314

Mendez, M. A. and Adair, L. S. (1999) 'Severity and timing of stunting in the first two years of life affect performance on cognitive tests in late childhood', *Journal of Nutrition*, vol 129, pp1555–1562

Morgan, K. J. (2004a) 'School meals and sustainable food chains: The role of creative public procurement', Caroline Walker Lecture, The Caroline Walker Trust, St Austell, UK

Morgan, K. J. (2004b) 'Sustainable regions: Governance, innovation and scale', *European Planning Studies*, vol 12, no 6, pp871–889

Morgan, K. J. (2006) 'School food and the public domain: The politics of the public plate', *The Political Quarterly*, vol 77, no 3, pp379–387

Morgan, K. J. (2007a) 'The polycentric state: New spaces of empowerment and engagement?' *Regional Studies*, vol 41, pp1237–1251

Morgan, K. J. (2007b) 'The ethical foodscape: Local and green versus global and fair', paper presented at the Economic and Social Research Council (ESRC) Science Week Conference on Local Food, St Asaph, UK, 13 March

Morgan, K. J. (2007c) 'Greening the realm: Sustainable food chains and the public plate', Centre for Business Relationships, Accountability, Sustainability and Society (BRASS) Working Paper Series No 43, Cardiff University, Cardiff

Morgan, K. J. and Morley, A. (2002) 'Relocalizing the food chain: The role of creative public procurement', the Regeneration Institute, Cardiff University, Cardiff

Morgan, K. J. and Morley, A. (2006) *Sustainable Public Procurement: From Good*

Intentions to Good Practice, Welsh Local Government Association, Cardiff

Morgan, K. J. and Sayer, A. (1988) *Microcircuits of Capital: Sunrise Industry and Uneven Development*, Polity Press, Cambridge, UK

Morgan, K. J. and Sonnino, R. (2005) *Catering for Sustainability: The Creative Procurement of School Meals in Italy and the UK*, the Regeneration Institute, Cardiff University, Cardiff

Morgan, K. J. and Sonnino, R. (2007) 'Empowering consumers: Creative procurement and school meals in Italy and the UK', *International Journal of Consumer Studies*, vol 31, no 1, pp19–25

Morgan, K. J., Marsden, T. K. and Murdoch, J. (2006) *Worlds of Food: Place, Power and Provenance in the Food Chain*, Oxford University Press, Oxford, UK

Morgan, K. J., Bastia, T. and Kanemasu, Y. (2007a) *Home Grown: The New Era of School Feeding*, World Food Programme, Rome

Morgan, K. J., Bastia, T. and Nicol, P. (2007b) *Nurturing Knowledge: An Evaluation of the Food Matters Project*, the Regeneration Institute, Cardiff University, Cardiff

Morris, C. and Young, C. (2000) '"Seed to shelf", "teat to table", "barley to beer" and "womb to tomb": Discourses of food quality and quality assurance schemes in the UK', *Journal of Rural Studies*, vol 16, pp103–115

Murdoch, J., Marsden, T. and Banks, J. (2000) 'Quality, nature and embeddedness: Some theoretical considerations in the context of the food sector', *Economic Geography*, vol 76, no 2, pp107–125

Murray, S. (2007) 'The deep fried truth', *The New York Times*, 14 December

Naselli, E. (2006) 'Meno carne e fritti, Italiani più salutisti studiano le etichette e temono gli OGM' ['Less meat and fried foods, more health-conscious Italians read the labels and fear GMOs'], *La Repubblica*, 24 October

National Audit Office (2005) *Sustainable Procurement in Central Government*, National Audit Office, London

National Audit Office (2006) *Ministry of Defence: Major Projects Report 200*, National Audit Office, London

Nelson, K. (1981) 'The school nutrition programs – Legislation, organization and operation', in *The National Evaluation of School Nutrition Programs. Review of Research* (Volume 1), System Development Corporation, Santa Monica, CA, pp27–100

Nestle, M. (2002) *Food Politics: How the Food Industry Influenced Nutrition and Health*, University of California Press, Berkeley, CA

Nestle, M. (2006) 'Food marketing and childhood obesity: A matter of policy', *The New England Journal of Medicine*, vol 354, pp2527–2529

O'Hara, S. U. and Stagl, S. (2001) 'Global food markets and their local alternatives: A socio-ecological economic perspective', *Population and Environment*, vol 22, no 6, pp533–554

Orrey, J. (2003) *The Dinner Lady*, Transworld, London

Page, L. (2006) *Lions, Donkeys and Dinosaurs*, Heinemann, London

Passmore, S. and Harris, G. (2004) 'Education, health and school meals: A review of policy changes in England and Wales over the last century', *Nutrition Bulletin*, vol 29, pp221–227

Paterson, M. (2000) *Understanding Global Environmental Politics: Domination, Accumulation and Resistance*, Macmillan, London

Petrini, C. (2001) *Slow Food: The Case for Taste*, Columbia University Press, New York

Pike, J. (2007) 'Lunchtime spaces', paper presented at the second Annual Conference, 'Free Healthy School Meals: The Hull Experience', Hull, UK, 17 November

Pirani, M. (2006) 'Il modello Roma va bene per l'Italia' ['The Roman model is good for Italy'], *La Repubblica*, 6 March

Policy Commission on the Future of Farming and Food (2002) *Farming and Food: A Sustainable Future*, Cabinet Office, London
Pollan, M. (2007) 'You are what you grow', *New York Times*, 22 April
Pomerantz, P. R. (2004) *Aid Effectiveness in Africa: Developing Trust between Donors and Governments*, Lexington Books, Lanham, MD
Poppendieck, J. (1998) *Sweet Charity? Emergency Food and the End of Entitlement*, Viking, New York
Poppendieck, J. (2008) *Stepping Up to the Plate: Realizing the Potential of School Food in America*, University of California Press, Berkeley, CA (forthcoming)
Pothukuchi, K. (2004) 'Community food assessment: A first step in planning for community food security', *Journal of Planning Education and Research*, vol 23, pp356–377
Pothukuchi, K and Kaufman, J. (2000) 'The food system: A stranger to urban planning', *Journal of the American Planning Association*, vol 66, no 2, pp113–124
Pretty, J. (1999) 'Reducing the costs of modern agriculture: Towards sustainable food and farming systems', in A. K. Dragun and C. Tisdell (eds) *Sustainable Agriculture and Environment: Globalisation and the Impact of Trade Liberalisation*, Edward Elgar, Cheltenham, UK
Pretty, J., Ball, A., Lang, T. and Morison, J. (2005) 'Farm costs and food miles: An assessment of the full cost of the UK weekly food basket', *Food Policy*, vol 30, no 1, pp1–20
Redclift, M. (1997) 'Frontiers of consumption: Sustainable rural economies and societies in the next century?', in H. DeHaan, B. Kasimis and M. Redclift (eds) *Sustainable Rural Development*, Ashgate, Aldershot, UK
Renard, M. (2003) 'Fair trade: Quality, markets and conventions', *Journal of Rural Studies*, vol 19, pp87–96
Renard, M. (2005) 'Quality certification, regulation and power in fair trade', *Journal of Rural Studies*, vol 21, pp419–431
Renting, H., Marsden, T. K. and Banks, J. (2003) 'Understanding alternative food networks: Exploring the role of short food supply chains in rural development', *Environment and Planning A*, vol 35, pp393–411
Renton, A. (2007) 'The great food aid con', *The Observer Food Monthly*, May
Richardson, D. (1997) 'The politics of sustainable development', in S. Baker et al (eds) *The Politics of Sustainable Development: Theory, Policy and Practice within the European Union*, Routledge, London
Roberts, C. (2008) *Healthy School Meals in Greenwich*, Greenwich Council, London
Ruffolo, U. (2001) 'Ristorazione scolastica: Prospettive giuridiche' ['School meals: Legal perspectives'], in R. B. Finocchiaro (ed) *La Ristorazione Scolastica: Prospettive Future* [*School Meals: Future Perspectives*], Quaderno 5, Cultura che Nutre, Programma Interregionale di Comunicazione ed Educazione Alimentare, Ministero delle Politiche Agricole e Forestali, Rome
Sachs, J. (2005) *The End of Poverty: Economic Possibilities for Our Time*, Penguin Press, New York
Sachs, W. and Santarius, T. (2007) *Slow Trade – Sound Farming: A Multilateral Framework for Sustainable Markets in Agriculture*, Heinrich Böll Foundation, Berlin
Sanchez, P., Swaminathan, M. S., Dobie, P. and Yuksel, N. (2005) *Halving Hunger: It Can Be Done*, UN Millennium Project Task Force on Hunger, Earthscan, London
Sassatelli, R. and Scott, A. (2001) 'Novel food, new markets and trust regimes: Responses to the erosion of consumers' confidence in Austria, Italy and the UK', *European Societies*, vol 3, no 2, pp213–244

Save the Children (2007) *Everybody's Business, Nobody's Responsibility: How the UK Government and the European Commission are Failing to Tackle Malnutrition*, Save the Children UK, London

Sayer, A. (2000) 'Moral economy and political economy', *Studies in Political Economy*, vol 62, pp79–104

Sayer, A. (2007) 'Moral economy as critique', *New Political Economy*, vol 12, no 2, pp261–270

Schibsted, E. (2005) 'Brain food: Nutritious eats + yummy ingredients = happy students', *Edutopia*, the George Lucas Educational Foundation, December, www.edutopia.org/brain-food

Schlosser, E. (2006) *Chew on This*, Puffin, London

School Food Trust (2006) *School Meal Provision in England and other Western Countries: A Review*, School Food Trust, London

School Meals Review Panel (2005) *Turning the Tables: Transforming School Food*, School Meals Review Panel, London

SCN (Standing Committee on Nutrition) (2006) 'Tackling the double burden of malnutrition: A global agenda', *Standing Committee on Nutrition News*, no 32, UN, Rome

Scottish Executive (2002) *Hungry for Success: A Whole School Approach to School Meals in Scotland*, The Stationery Office, Edinburgh

Scottish Executive (2004) 'Integrating sustainable development into public procurement of food and catering services', www.scotland.gov.uk/Resource/Doc/1265/0005191.pdf

Scottish Parliament (2007) *Official Report: Green Procurement*, Scottish Parliament, Edinburgh, 1 February

Severson, K. (2007) 'Local carrots with a side of red tape', *The New York Times*, 17 October

Seyfang, G. (2006) 'Ecological citizenship and sustainable consumption: Examining local organic food networks', *Journal of Rural Studies*, vol 22, pp383–395

Sharp, I. (1992) *Nutritional Guidelines for School Meals*, Caroline Walker Trust, London

Sheeran, J. (2008) 'Opening statement to Executive Board', World Food Programme, 4 February, Rome

Sign (2006) *Signals Newsletter*, November, www.sign-schoolfeeding.org

Simon, M. (2006) *Appetite for Profit: How the Food Industry Undermines our Health and How to Fight Back*, Nation Books, New York

Simpson, A. (2006) 'Buy in Africa, sustain local communities', *International Trade Forum*, no 3

Smargiassi, M. (2007) 'Mense d'Italia: Tre milioni a tavola' ['Canteens of Italy: Three million people at the table'], *La Repubblica*, 18 October

Smith, A (2006) 'Council scraps free school meals', *The Guardian*, 7 June

Smith, D. (1998) How far should we care? On the spatial scope of beneficence', *Progress in Human Geography*, vol 22, no 1, pp15–38

Smithers, R. (2008) 'Ban junk food advertising on internet, say campaigners', *The Guardian*, 15 March

Sneddon, C., Howarth, R. B. and Norgaard, R. B. (2006) 'Sustainable development in a post-Brundtland world', *Ecological Economics*, vol 57, pp253–268

Soil Association (2003) *Food for Life: Healthy, Local, Organic School Meals*, Soil Association, Bristol, UK

Sonnino, R. (2007a) 'Quality for all: School meals in Rome', in V. Wheelock (ed) *Healthy Eating in Schools*, Verner Wheelock Associates, Skipton, UK, pp81–190

Sonnino, R. (2007b) 'Embeddedness in action: Saffron and the making of the local in southern Tuscany', *Agriculture and Human Values*, vol 24, pp61–74

Sonnino, R. (2007c) 'United Kingdom: A desk review of East Ayrshire's local school meals', in K. J. Morgan et al (eds) *Home-Grown: The New Generation of School Feeding*, World Food Programme, Rome

Sonnino, R. (2009) 'Quality food, public procurement and sustainable development: The school meal revolution in Rome', *Environment and Planning A*, in press

Sonnino, R. and Marsden, T. (2006) 'Beyond the divide: Rethinking relationships between alternative and conventional food networks in Europe', *Journal of Economic Geography*, vol 6, pp181–199

Sonnino, R. and Morgan, K. J. (2007) 'Localizing the economy: The untapped potential of green procurement', in A. Cumbers and G. Whittam (eds) *Reclaiming the Economy: Alternatives to Market Fundamentalism in Scotland and Beyond*, Scottish Left Review Press, Biggar, UK, pp127–140

South Gloucestershire Department for Children and Young People (2005) *Catering in Primary Schools*, South Gloucestershire Department for Children and Young People, Chipping Sodbury, UK

Spake, A. (2005) 'The world of Chef Jorge. It's a daunting task: Make New York City's school lunches healthful – and fun to eat', *US News and World Report*, vol 138, p64

Stern, N. (2006) *The Economics of Climate Change: The Stern Review*, HM Treasury, London

Sustain (2005) *Sustainable Food Procurement in London's Public Sector*, Sustain, London

Sustainable Procurement Task Force (2006) *Procuring the Future: Sustainable Procurement National Action Plan*, Defra, London

Tisdell, C. (1999a) 'Conditions for sustainable development: Weak and strong', in A. K. Dragun and C. Tisdell (eds) *Sustainable Agriculture and the Environment: Globalisation and the Impact of Trade Liberalisation*, Edward Elgar, Cheltenham, UK

Tisdell, C. (1999b) 'Economics, aspects of ecology and sustainable agricultural production', in A. K. Dragun and C. Tisdell (eds) *Sustainable Agriculture and the Environment: Globalisation and the Impact of Trade Liberalisation*, Edward Elgar, Cheltenham, UK

Thorpe, L. E., List, D., Marx, D. G., May, L., Helgerson, S. D. and Frieden, T. (2004) 'Childhood obesity in New York City elementary school students', *American Journal of Public Health*, vol 94, pp1496–1500

Travers, T. (2004) *The Politics of London: Governing the Ungovernable City*, Palgrave, Basingstoke, UK

Tronto, J. (1993) *Moral Boundaries: A Political Argument for an Ethic of Care*, Routledge, New York

United Nations (2007) 'Working together to end child hunger and undernutrition', *Standing Committee on Nutrition News*, no 34, UN, Rome

United States Department of Health and Human Services (2001) 'The Surgeon General's call to action to prevent and decrease overweight and obesity', United States Department of Health and Human Services, Rockville, MD

USDA Food and Nutrition Service (2005) 'Eat smart – Farm fresh! A guide to buying and serving locally-grown produce in school meals', www.fns.sda.gov/cnd/Guidance/Farm-to-School-Guidance_12-19-2005.pdf

Vallianatos, M., Gottlieb, R. and Haase, M. (2004) 'Farm to school: Strategies for urban health, combating sprawl and establishing a community food systems approach', *Journal of Planning Education and Research*, vol 23, pp414–423

Van Egmond-Pannell, D. (1985) *School Foodservice*, AVI Publishing Company, Westport, CT

Veltroni, W. (2006) 'Il modello Roma' ['The Roman model'], *Italianieuropei*, January/February, pp143–148

VITA Non-Profit Online (2003) 'Bio: Biologiche due mense su tre' ['Organics: Two canteens out of three are organic'], www.vita.it/articolo/index.Php3?NEWSID=35208

von Braun, J. (2007) *The World Food Situation: New Driving Forces and Required Action*, International Food Policy Research Institute (IFPRI), Washington, DC

von Braun, J. (2008) *Food Prices, Biofuels, and Climate Change*, International Food Policy Research Institute (IFPRI), Washington, DC

von Schirnding, Y. (2002) 'Health and sustainable development: Can we rise to the challenge?', *The Lancet*, vol 360, pp632–637

Watts, D. C. H., Ilbery, B and Maye, D. (2005) 'Making reconnections in agro-food geography: Alternative systems of food provision', *Progress in Human Geography*, vol 29, pp22–40

WCED (World Commission on Environment and Development) (1987) *Our Common Future*, Oxford University Press, Oxford, UK

Wekerle, G. R. (2004) 'Food justice movements: Policy, planning and networks', *Journal of Planning Education and Research*, vol 23, pp378–386

Welsh Procurement Initiative (2005) *Food for Thought – A New Approach to Public Sector Food Procurement: Case Studies*, Welsh Procurement Initiative, Cardiff

WFP (World Food Programme) (2004) *Global School Feeding Report*, World Food Programme, Rome

WFP (2005) 'Country programme, Ghana: 2006–2010', Executive Board Document, World Food Programme, 6 September

WFP (2006) *Global School Feeding Report*, World Food Programme, Rome

WFP (2007a) 'Support to Nepad: Period of report: 2003–2004', available at www.un.org/africa/osaa/cpcreports/28.WFP_formatted.pdf/

WFP (2007b) *Home-Grown School Feeding Field Case Study: Ghana*, World Food Programme, Rome

WFP (2008) 'Information sheet on procurement', World Food Programme, Rome

Which? (2006) 'Child catchers: The tricks used to push unhealthy food to your children', *Which?*, London

WHO (2004) *Marketing Food to Children: The Global Regulatory Environment*, World Health Organization, Geneva

Williamson, E. (2008) 'Bad beef in schools', www.chefann.com/blog?p=980

Williamson, H. (2007) 'West warned of anti-graft drive', *Financial Times*, 16 November

Winter, M. (2003) 'Embeddedness, the new food economy and defensive localism', *Journal of Rural Studies*, vol 19, pp23–32

World Bank (2006) *Repositioning Nutrition as Central to Development*, World Bank, Washington, DC

Wright, S. (2007) 'New York City school dinners', *Caterer and Hotelkeeper*, 18 January

Wrigley, N. (2002) '"Food deserts" in British cities: Policy context and research priorities', *Urban Studies*, vol 39, no 11, pp2029–2040

WTO (2007) *Understanding the WTO: The Agreements*, World Trade Organization, Geneva

Index